WATER WITHOUT BORDERS?

Canada, the United States, and Shared Waters

Since 1909, the waters along the Canada-US border have been governed in accordance with the Boundary Waters Treaty, but much has changed in the last 100 years. This engaging volume brings together experts from both sides of the border to examine the changing relationship between Canada and the United States with respect to shared waters, as well as the implications of these changes for geopolitics and the environment. *Water without Borders?* is a timely publication given the increased attention to shared water issues, and particularly because 2013 is the United Nations International Year of Water Cooperation.

Water without Borders? is designed to help readers develop a balanced understanding of the most pressing shared water issues between Canada and the United States. The contributors explore possible frictions between governance institutions and contemporary management issues, illustrated through analyses of five specific transboundary water "flashpoints." The volume offers both a historical survey of transboundary governance mechanisms and a forward-looking assessment of new models of governance that will allow us to manage water wisely in the future.

EMMA S. NORMAN is an assistant professor of Geography at Michigan Technological University.

ALICE COHEN is an assistant professor in Environmental Science and Environmental and Sustainability Studies at Acadia University.

KAREN BAKKER is Canada Research Chair in Political Ecology, director of the Program on Water Governance, and a professor in the Department of Geography at the University of British Columbia.

Water without Borders?

Canada, the United States, and Shared Waters

EDITED BY EMMA S. NORMAN, ALICE COHEN, AND KAREN BAKKER

UNIVERSITY OF TORONTO PRESS
Toronto Buffalo London

© University of Toronto Press 2013
Toronto Buffalo London
www.utppublishing.com
Printed in Canada

ISBN 978-1-4426-4393-2 (cloth)
ISBN 978-1-4426-1237-2 (paper)

Printed on acid-free, 100% post-consumer recycled paper with vegetable-based inks.

Publication cataloguing information is available from Library and Archives Canada

This book has been published with the help of a grant from the Canadian Federation
for the Humanities and Social Sciences, through the Awards to Scholarly Publications
Program, using funds provided by the Social Sciences and Humanities Research
Council of Canada.

Financial support from the Walter and Duncan Gordon Foundation is gratefully
acknowledged.

University of Toronto Press acknowledges the financial assistance to its publishing
program of the Canada Council for the Arts and the Ontario Arts Council.

University of Toronto Press acknowledges the financial support of the Government of
Canada through the Canada Book Fund for its publishing activities.

All royalties from the sale of this book go to the Canada and US branches of the
Waterkeeper Alliance.

Contents

List of Tables

List of Figures

List of Boxes

Acknowledgments

All royalties from Water without Borders? *will be donated to the Canadian and US branches of Waterkeepers Alliance – "a global movement of on-the-water advocates who patrol and protect over 100,000 miles of rivers, streams and coastlines in North and South America, Europe, Australia, Asia and Africa"* (www.waterkeeper.org*).*

We hold up our (water) glasses in appreciation for all of those in the trenches – particularly the authors of this volume – who continue to show us how problems can inspire innovative solutions in the governance of shared waters.

The success of this project relied on the support of many individuals.

We would like to thank the University of Toronto Press for their exceptional professionalism and guidance throughout this process. We would also like to thank our editor, Daniel Quinlan, who has shown unwavering enthusiasm for and support of this project throughout its duration. Thanks also to Apex CoVantage for superb copyediting; Shoshona Wasser, Deepshikha Dutta, and Brian MacDonald for their help with promotion and advertising; and the three anonymous reviewers for their keen insights and excellent suggestions.

Financial support from the Walter and Duncan Gordon Foundation (http://www.gordonfoundation.ca) for the authors' workshop, the book's website, and promotion and dissemination of this project is gratefully acknowledged.

At The University of British Columbia (UBC), we would like to thank Eric Leinberger for his excellent cartographic skills and Sandy Lapsky for providing sustained administrative support. At the Program on Water Governance, we gratefully acknowledge the support of Gemma Dunn, Leila Harris, Christina Cook, Corin de Frietas, and Hana Galal, who helped organize the authors' workshop. Thanks also to Brittany Flaherty who helped with copyediting and who contributed Box 1.2, "The Boundary Waters Treaty: The Next 100 Years." In-kind support from UBC's Institute for Resources, Environment, and Sustainability was instrumental to the authors' workshop.

Also, thanks to Sigrid Emrich, David Hermann, Patrick Higgins, Dean Sherratt, and David Whorley for attending the authors' workshop and providing valuable commentary, and the Consulate General of Seattle for hosting the workshop's reception.

Terry Drussel (Fat Black Squirrel Studio) designed the Water without Borders logo, Harvey Locke allowed us to use his inspiring Flathead Valley photographs, Kyle Robertson designed our website (www.waterwithoutborders.info), and the San Francisco – based Stamen Design Group provided the open-source cover image (www.stamen.org).

Last, but not least, we acknowledge our families, who continue to lovingly support our endeavours. To Chad, Parker, Luke, Jamie, Philippe, Solenne, and Pauline.

Emma Norman
Houghton, Michigan

Alice Cohen
Wolfville, Nova Scotia

Karen Bakker
Vancouver, British Columbia

WATER WITHOUT BORDERS?

Canada, the United States, and Shared Waters

1 Introduction

EMMA S. NORMAN, ALICE COHEN, AND KAREN BAKKER

The title of this book asks readers to envision the possibility of water without borders. This vision may seem obvious: Water is already "without borders" in the sense that it is continually flowing across boundaries of all kinds – political, ecological, social, and economic. Yet borders remain crucial for water, for two reasons. First, the (geo)political borders embedded in contemporary water management systems have significant – and often negative – effects on watercourses and water users. Second, the reality of global water challenges and their local effects highlights the need for innovative strategies that are locally appropriate, globally coordinated, ecologically sensitive, and politically feasible – a combination that requires strategic negotiation of political borders to safeguard this most precious and essential substance.

Conceptualizing water without borders, then, may be a first step in addressing one of the world's most pressing issues – ensuring access to water of adequate quantity and quality for all of humanity, while preserving water for nonhuman uses and ecosystem services (Vösörmarty et al. 2010). Achieving this goal requires cooperation across multiple scales and communities, from international governance organizations to neighbourhoods. A particular challenge is the issue of transboundary freshwater:[1] currently, at least 263 international waterways (lakes and rivers) and countless groundwater sources span international borders. Almost two-thirds of states with international water Basins have at least half of their territory in an international basin, and the remaining third has more than 80 per cent in an international territory (Conca 2006). Mismatches between the hydrologic cycle and administrative borders is thus commonplace.

National governments have developed various mechanisms to address this mismatch. At least 145 binational and multilateral water-related treaties have been created worldwide since 1814.[2] Although "water wars" are much discussed, the historical record demonstrates that these wars are, in fact, a rare occurrence.[3] As geographer Aaron Wolf (2001) argues, incentives for cooperation usually outweigh those for conflict. In fact, scholars argue, water is more than twice as likely to be a source of international cooperation as of conflict (Wolf 2001; see also Gleick 2007).

This book examines the transboundary water relationship between Canada and the United States within this broader context. Our goal is to highlight best practices and discuss areas of continued concern relevant to international and domestic audiences. This goal is particularly important given recent strains in the shared governance of waters across the Canada-US border. While addressing these issues, we also explore how the Canada-US border offers a worthy example of effective transboundary water governance. These two countries have worked together for more than 100 years – through changing economic and social climates – to co-manage shared resources. This co-management model has evolved, shaped partly by geography and partly by geopolitical and historical circumstance. With more than 8,800 kilometres (5,468 miles) of shared borders (including 2,475 kilometres [1,358 miles] with Alaska), and some of the world's largest bodies of freshwater to co-manage, the potential for Canada-US conflict, as well as cooperation, is considerable. Interestingly, the geography of river basins along the Canada-US border is such that most large basins – for example, the McKenzie River Basin – fall neatly onto one side or the other of the border; exceptions are the Great Lakes Basin and the Columbia River Basin. The location of these basins, as pointed out in chapter 5, was a major consideration of negotiators when they settled the international boundary at the 49th parallel.

This book explores the mechanisms put in place more than a century ago. These mechanisms created a structured set of shared governance and conflict-resolution processes designed to minimize or eliminate conflict. Arguably, these mechanisms have worked relatively well since their inception (albeit with several long-standing exceptions). A central question we posed to contributing authors, then, is whether and how existing governance relationships need to evolve in light of changes to the Canada-US relationship, new approaches to environmental governance, and greater pressures on – and increasingly unpredictable availability of – water resources, both regionally and globally (UNW-WAP 2012; Wouters and Ziganshina 2011; Vörösmarty et al. 2000, 2010).

A Century of Evolving Water Governance along the Border

One of the most important actors in transboundary Canada-US water governance is the International Joint Commission (IJC), whose history and role is briefly explained here; it is revisited in more detail in subsequent chapters (particularly chapter 4 and boxes 1.1 and 1.2). In the early 1900s, Canada recognized that its southern neighbour was emerging as an international powerhouse; Canada was keen to negotiate a treaty to preserve its water rights, create a mechanism for cooperation in transboundary, and establish fair arbitration for (what were anticipated to be unavoidable) water disputes. Against this backdrop, the British ambassador in Washington, DC (on Canada's behalf) drafted a treaty with the US government. The resulting Boundary Waters Treaty (BWT) has become one of the longest-standing water-related treaties in the world (Heinmiller 2008).

Box 1.1   The International Joint Commission

Murray Clamen

"I do not anticipate that the time will ever come when this Commission will not be needed. I think that as the two countries along this tremendous boundary become more and more thickly settled the need for it will increase."
— Elihu Root, Secretary of State, 1909 (Munton 1981)

The 1909 Boundary Waters Treaty provided for the establishment of the International Joint Commission (IJC), which held its first meeting in Washington, DC on 10 January 1912.

From the beginning, the IJC's fundamental role has been to prevent and resolve transboundary water disputes between Canada and the United States under the Boundary Waters Treaty (BWT). Before proceeding with their work, each of the six commissioners (three representing Canada and three representing the United States) make a solemn declaration to perform their duties faithfully and impartially. Commissioners act as a single unit, seeking consensus solutions to common problems in the joint interest of both countries. In the conduct of their duties, commissioners are supported by staffs in the two section offices (Secretariats) in Ottawa and Washington, a Great Lakes Regional Office in Windsor, Ontario, as well as some 20 expert advisory boards and ad hoc task forces whose members are appointed by the commission. Board members, like commissioners, are instructed to act at all times in their "personal and professional capacity," and not as representatives of their employing organizations or departments.

The specific activities of the IJC are as follows:

- Under Articles III and IV of the BWT, the IJC acts on applications for hydropower dams and other projects along the border to protect all interests from the effects of such projects.
- Under Article IX, the two governments may refer questions or matters of difference to the IJC for examination and report. When the IJC receives such a reference, it usually appoints an investigative board or task force to examine the facts and advise on the governments' questions. The IJC's recommendations are not legally binding, but are almost always taken up.
- Under Article X of the BWT, governments may refer any issue to the commission for binding decision rather than for report and recommendation. However, this function has never been used in the over 100-year histories of the BWT and the IJC.

Although the IJC has been very successful in mitigating Canada-US transboundary water issues, the Commission is facing a number of changes, including declining references (see chapter 3) and the development of international watershed boards (see chapter 4). In the face of such changes, what will become of the IJC over the next 100 years?

Reference

Munton, D. 1981. "Dependence and Interdependence in Transboundary Environmental Relations." *International Journal* 36 (1): 139–84.

Box 1.2   The Boundary Waters Treaty: The Next 100 Years

Brittany Flaherty

At the end of the nineteenth century, a handful of disputes had emerged between the United States and Canada over the Niagara, Rainy, St. Mary, and Milk Rivers (IJC 2011a). In light of these disputes, the Canadian government in 1896 made clear its readiness to cooperate in regulation of international streams. In 1902, the United States moved forward with this suggestion and invited Canada to form a joint international commission to investigate and report on waters the United States and Canada shared. The resulting International Waterways Commission experienced only limited success in having its recommendations implemented, but did lead to negotiations for a new treaty to govern the use of boundary waters (IJC 2011a). On 11 January 1909, the Boundary Waters Treaty (BWT) was signed by Secretary of State Elihu Root and Ambassador James Bryce in Washington to ensure the equitable use of shared waters, providing "the principles and mechanisms to help resolve disputes and to prevent future ones" (IJC 2011b).

Today, the treaty is considered a great success (Fogarty et al. 2010; Hall 2008; Knox 2008). The International Joint Commission, established under the BWT, receives credit for having "effectively and peacefully managed the boundary waters of two nations over some ninety years, reconciling or averting more than 130 disputes" (Wolf 2001, 33).

Yet some of the most important emerging concerns were simply not on the radar of those developing the BWT over a century ago. The BWT was intended to ensure freedom of navigation and regulation of diversions, and to control transboundary pollution – issues that were the primary transboundary concerns of the early twentieth century. The BWT's goal was to "allow Canada and the United States to use their boundary waters in ways that would not unduly interfere with one another, not to ensure that the ecosystems they shared remained healthy" (Knox 2008). Environmental factors such as ecosystem protection and groundwater consequently are not mentioned whatsoever in the treaty, and Aboriginal rights are omitted altogether. These oversights are at the heart of many of the "flashpoints" this book discusses. Indeed, despite the many changes and challenges of the past century, the BWT has remained the same, never having been altered or amended in any way (Hall 2008). Critics often call into question the treaty's effectiveness in handling modern disputes and challenges. This lack of effectiveness is reflected in the declining references discussed in chapter 3 of this volume and exemplified in the Devils Lake controversy outlined in chapter 11.

The chapters that follow show how actors work through, and around, the provisions of a century-old treaty. The central question as the United States and Canada move forward is: what will the next 100 years look like?

References

Fogarty, K., et al. 2010. "Emerging Legal Issues in the Great Lakes Such as the Public Trust Doctrine, Subterranean Rights and Municipal Regulatory Arrangements." *Canada-United States Law Journal* 34 (2): 279–320.

Hall, N.D. 2008. "Centennial of the Boundary Waters Treaty: A Century of United States-Canadian Transboundary." *Wayne Law Review* 54 (4): 1417–50.

IJC. 2011a. "Origins of the Boundaries Water Treaty." International Joint Commission. http://bwt.ijc.org/index.php?page=origins-of-the-boundaries-water-treaty&hl=eng. Accessed 24 October 2011.

IJC. 2011b. "Treaties and Agreements." International Joint Commission. http://www.ijc.org/rel/agree/water.html. Accessed 25 October 2011.

Knox, J.H. 2008. "The Boundary Waters Treaty: Ahead of its Time and Ours." *Wayne Law Review* 54:1591–608.

Wolf, A.T. 2001. Transboundary Waters: Sharing Benefits, Lessons Learned. International Conference on Freshwater-Bonn. http://www.agnos-online.de/inwent1/images/pdfs/transboundary-waters.pdf. Accessed 28 September 2011.

Box 1.3   Articles of the Boundary Waters Treaty

Preliminary Article – Boundary waters are defined as "the lakes and rivers and connecting waterways ... along which the international boundary between the United States and the Dominion of Canada passes ... but not including tributary waters."

Article I – Ensures that the navigation of all boundary waters will continue, free and open to both countries equally, but subject to laws and regulations of either country within its own territory. These same rights of navigation are extended to Lake Michigan and all canals connecting boundary waters, now existing or which are later constructed on either side of the line.

Article II – Each side of the border has exclusive jurisdiction and control over the use and diversion of all waters on its own side. However, any interference with, or diversion of, these waters that results in injury on the other side of the boundary guarantees those parties the same legal remedies as if the injury took place in the country of the diversion or interference.

Article III – No obstructions or diversions that affect the natural level or flow of boundary waters on either side of the line shall be made, except by authority of the United States or Canada within their respective

Figure 1.1 The first meeting of the International Joint Commission, 1912.

International Joint Commission – Jany 1912.

jurisdictions and with the approval of the International Joint Commission (see Figure 1.1).

Article IV – The two nations will not permit the construction or maintenance on their respective sides of any works or obstructions that raise the natural water level on the other side of the boundary, unless approved by the International Joint Commission. Boundary and transboundary waters will not be polluted on either side to the injury of health or property on the other.

Article V – No diversion of the waters of the Niagara River above the Falls will be permitted.

Article VI – The St. Mary and Milk Rivers and their tributaries are to be treated as one stream for the purpose of irrigation and power, and the waters will be apportioned equally between the two countries, but more or less than half can be taken by each to afford more beneficial use to each.

Article VII – The parties agree to establish and maintain an International Joint Commission of the United States and Canada composed of six commissioners, three appointed by each nation.

Article VIII – This commission will have jurisdiction over and pass upon all cases involving the use, obstruction, or diversion of the waters with the following uses in order of precedence: domestic and sanitary purposes, navigation, and power and irrigation purposes.

Article IX – Any questions or differences arising between the parties may be referred to the commission for examination, whenever one nation refers the issue.

Article X – Any questions or differences may be referred to the commission for decision by consent of the two parties. A majority of the commission will have the power to render a decision. If equally divided, the IJC must make a joint report to the parties.

Article XI – All decisions and joint reports made by the commission will be filed with the secretary of state of the United States and the governor general of the Dominion of Canada.

Article XII – The International Joint Commission will meet and organize at Washington promptly after the members are appointed and arrange times and places for its meetings.

Article XIII – All previously referenced special agreements between the involved parties referred to in the above articles are understood also to include mutual arrangements expressed by concurrent or reciprocal legislation on the part of Congress and the Parliament of the Dominion.

Article XIV – This treaty will be ratified by the president of the United States, with the advice and consent of the senate, and by His Britannic Majesty in Washington as soon as possible.

Reference

International Joint Commission (IJC) United States and Canada. 1909. "Who We Are." http://www.ijc.org/rel/agree/water.html#text

The BWT mandated the creation of a governing body, the IJC, which investigates transboundary water disputes, provides recommendations to the Canadian and US governments, and monitors ongoing hydraulic operations (dams and diversions) at the border. The IJC, a binational model of international relations, relies on cooperation between federal agencies in Ottawa and Washington, DC . This federal-to-federal model was typical of international relations in the first half of the twentieth century, making it the structure of choice for decision-makers crafting the BWT and the subsequent IJC. This "federal era" governance model corresponded with the development of large-scale hydraulic projects (for example, the regulation of the Columbia River and the St. Lawrence Seaway) (Pentland and Hurley 2007). Table 1.1 shows the evolution of Canada-US transboundary management.

Table 1.1. Eras of Canada-US Transboundary Water Management (1945–Present)

Transboundary Water Era	Time Period	Role	Example
Cooperative Development	1945–1965	• Projects of mutual benefit • Federal government – encouraged hydroelectric development	Columbia River Treaty; St. Lawrence Seaway
Comprehensive Management	1965–1985	• Issue based • Comprehensive river basin planning and more "environmentally conscious" framework • Water expertise built up at federal level	Great Lakes Water Quality Agreement
Sustainable Development	1985–2000	• Link economy and environment • Issues more integrative, anticipatory and preventive	Great Lakes Annex
Participatory	2000–current	• Increased local participation	Watershed Boards

Sources: Adapted from Pentland and Hurley (2007) and Norman and Bakker (2009).

As Table 1.1 shows, the old model of federal-to-federal agreements is changing. Since the early 1980s, a rescaling of governance has occurred in the management of natural resources, including water. Now, rather than only federal-to-federal negotiations, the actors involved in decision-making processes represent different jurisdictional scales (federal, provincial, state, local, and First Nation and Native American communities). This representation creates ever-increasing complexity in governance, which is one of the central issues subsequent chapters explore. Moreover, an increasing number of individual citizens, nongovernmental organizations, and private enterprises participate in decision-making processes. In light of this evidence, the scholarly literature suggests that two simultaneous rescaling processes are at work: (1) a scaling *out* from government, toward increased public participation, and (2) and a scaling *down* and up from higher orders of government toward watersheds (which also often constitutes a scaling up from municipalities) (Cohen 2012; Reed and Brunyeel 2010). Environmental governance – including water governance – has also moved toward more integrated and holistic approaches that endeavour to overcome gaps and overlaps between academic disciplines, government departments, and jurisdictional responsibilities. In the case of water governance, this has taken the form of integrated water resource management (Mitchell 1990), source water protection (Cook 2011), and watershed-scale governance (Cohen 2012).

These changes to environmental governance are overlaid on long-standing transboundary water arrangements and contentious water bodies (which we term "flashpoints' or "hotspots"; see Figure 1.2). In some instances, conflict between users has led to the development of models of collaboration and cooperation, such as those found in the Great Lakes. But, in recent years, new conflicts have arisen and dormant conflicts have reignited. Accordingly, commentators increasingly question the adequacy and relevance of the transboundary water governance framework the BWT created over a century ago.

We define water governance as the range of political, organizational, and administrative processes through which communities articulate their interests, their input is absorbed, decisions are made and implemented, and decision-makers are held accountable in the development and management of water resources and delivery of water services (Bakker 2007; Nowlan and Bakker 2010). The key point here is that we define water governance as involving more than laws: governance is a continuously evolving decision-making process, including meetings, reports, data collection, public feedback, decisions, elections, and so on. Water governance scholarship focuses on these processes at all scales: international (Conca 2006; European Union 2011; United Nations 2011),

Figure 1.2 Flashpoints along the Canada-US international border.

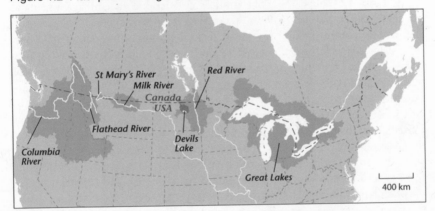

Source: Original Map by Eric Leinberger, University of British Columbia.

national (Bakker 2007; Barlow 2007; Deason, Schad, and Sherk 2001), as well as subnational level, particularly regarding state, provincial, and local strategies (Hill et al. 2008; Norman 2009; Gooch and Rieu-Clarke 2010; Forest 2011). In addition, the strengthening and refining of Aboriginal rights have created a "third sovereign" acting in collaboration with existing frameworks (Hele 2008; Norman 2012, 2013; Phare 2009; Simpson 2000, 2007; Thom 2010).

Borders of All Kinds: Political, Hydrologic, and Social

Much of this book's analysis focuses on one international political border: the forty-ninth parallel, which forms the international boundary between Canada and the United States (see Figure 1.3). This border is crossed – and at times made up of – the waters this book highlights.

Figure 1.3 The Canada-US international border cut looking east from the Flathead Valley.

Source: Harvey Locke.

Yet this is not the only boundary that affects the way in which water is governed, and these other boundaries bear mention here. The first additional set of boundaries is hydrologic: the topographical lines that define the flow of surface water, and the hydrogeogic factors that shape the flow of groundwater. These hydrologic boundaries are the subject of ever-increasing attention in the water governance literature.[4] The fact that hydrologic and political boundaries do not coincide[5] is noted frequently, and an increasing number of voices has advocated for the primacy of hydrologic boundaries for the purposes of water governance, prompting a call for a post-sovereign, hydrologically based approach to water governance (Budds and Hinojosa 2012). These discussions have led to nuanced interpretations of the interactions between society and water (often called the "hydro-social cycle")[6] and new interpretations of how to frame issues related to water governance, borders, and scale, such as transboundary inter-local water supplies (Conca 2008; Forest 2011). Although this book does not focus centrally on the question of hydrologic boundaries, we reference their increasing importance throughout the book: Norman and Bakker's chapter on rescaled governance and Clamen's discussion of the evolution of the IJC are particularly good examples.

The second additional set of boundaries relates to social and cultural boundaries. Linguistic boundaries within Canada are one example, as are the First Nations boundaries that Phare describes in chapter 2. Moreover, the lengthy border plays host to several cross-border regional identities: "Cascadia" and the "Salish Sea" in the Western regions, the transboundary prairie region, the "Freshwater Nation" of the Great Lakes, and the Atlantic region. Regional identities such as these can play a significant role in public imaginations and can influence transboundary relations far from national capitals.

These alternative borders remind us that boundaries of all kinds are in a constant state of construction and reconstruction through nuanced ecological, political, and social interactions. As explored throughout the book (and, particularly, chapters 2 and 3), borders are simultaneously socially and physically constructed – through politics, power struggles, and in many cases, the materiality of border infrastructure.

The Aims of *Water without Borders?*

This book answers the following three questions:

1. What are the most important trends in Canada-US transboundary water governance, focusing on the past century?

2. What are the key challenges and opportunities facing water managers, politicians, and water users?
3. How can Canada and the United States move forward with effective transboundary water governance?

In engaging with these questions, this book grapples with issues central to the practice of transboundary water governance. Posing these questions helps conceptualize "water without borders" – in other words, new forms of water governance that are not constrained by geopolitical boundaries.

The chapters in this volume each explore the challenges and solutions germane to their topical focus. Moreover, the authors use these broader debates to frame their work and query the extent to which the portrayal of Canada-US transboundary as a world model is mirrored in their work. In so doing, the book asks: Are the lessons from the Canada-US example transferable to other parts of the world? We believe that the answer is yes.

Yet despite the exemplary nature of Canada-US water relations, "flashpoints" (specific water bodies that have inspired disputes) continue to remind us of the importance of cultivating relationships and (re)creating good governance mechanisms. This reminder is particularly salient in the post-9/11 era, where federal funds continue to support tightening – rather than opening – the border (tightening the border is particularly pronounced on the US side).

The contributors to *Water without Borders?* help unpack the multifaceted issues facing governance of transboundary water. The volume is divided into two sections:

- Part I: Issues, Approaches, and Challenges
  This section situates the Canada-US transboundary experience within the context of changing hydrological landscapes. It flags key issues and identifies lessons that we can learn from the ongoing (and changing) relationship between actors governing water along the Canada-US border. This section outlines new models of transboundary water governance and highlights current work that looks beyond current governance frameworks to address pressing and timely issues.
- Part II: Flashpoints, Conflict, and Cooperation
  This section presents short commentaries on key transboundary "flashpoint" water bodies along the Canada-US border. For each flashpoint, a Canadian author and a US author were invited to co-write a chapter, working together to assess the key issues, including lessons learned from controversy and collaboration, and possible traps to avoid in the future.

In addition, the authors articulate how problems, in many cases, can inspire innovative solutions.[7]

Chapter Organization

Part I: Issues, Approaches, and Challenges

In the first part of *Water without Borders?*, the authors characterize those issues that impact the health and sustainability of our shared waters. Authors Phare (chapter 2), Norman and Bakker (chapter 3), and Clamen (chapter 4) each explore the changing face of transboundary water governance. In chapter 2, Phare sets the scene by analysing critical questions regarding the construction of borders in relation to flow resources such as water; Phare provides a nuanced interpretation of borders and water in a postcolonial context, deepening our understanding of changing governance patterns. The chapter's premise is that all political borders are both colonial and socially constructed, and that the impacts of these borders are particularly acute for Indigenous peoples, whose traditional territories often have been fragmented by externally determined and controlled lines. Phare highlights two cases where First Nations are actively working in water governance practices and defining the terms of governance for and by their communities.

In chapter 3, Norman and Bakker analyse trends in water governance along the Canada-US borderland over the past century. Most notably, changing patterns in environmental management practices have led to increased participation of local actors and greater public participation in governance processes. The authors find that increased participation of local actors does not equate to increased capacity or greater influence within the decision-making process. The authors then discuss strategies for increasing the institutional capacity of local actors to enable meaningful local participation in transboundary water governance.

In chapter 4, Clamen describes how the roles and responsibilities of the IJC have evolved over the 100-year existence of this governing body. Clamen notes how institutional flexibility has allowed the IJC to explore the possibility of moving from a response-based approach to specific issues to a more proactive approach using international watershed boards. As part of these discussions, the author looks at the shifting relationship of actors participating in the governance of shared waters, including the role of First Nations and Native Americans as a third sovereign, and analyses the importance of cross-scale influence and cooperation.

In chapter 5, Lasserre analyses one of the most controversial contemporary water issues facing Canada and the United States – continental bulk water transfers. Lasserre analyses these public debates and dispel some commonly held assumptions about the possibility of bulk water transfer between Canada and the United States. The chapter authors also provide a historical analysis of the emergence of water diversion schemes and ask the critical question: "Are continental bulk water transfers a chimera, or do they have real potential?" The also show that many bulk water transfers are already happening.

Chapter 6 provides a broad overview of key water management challenges currently facing Canada and the United States. Climate change, threats to groundwater, and the changing face of water pollution are some of the challenges that require immediate attention and sustained engagement. Understanding the key challenges Canada and the United States currently face regarding water governance helps set the scene for the discussions found in Part II: Flashpoints, Conflict, and Cooperation.

Both chapters 5 and 6 examine the potential for moving toward more lasting solutions by shifting priorities and turning these challenges into opportunities – a discussion that is continued in Part II.

Part II: Flashpoints, Conflict, and Cooperation

The chapters in Part II of the volume present short commentaries on key "flashpoints": Canada-US transboundary waters that are (or have the potential to be) sources of conflict. Each chapter, jointly authored by experts from either side of the international border, focuses on a single flashpoint. The Canadian and US co-authors collaborated on writing each chapter, providing a nuanced perspective of each flashpoint and a collaborative assessment of key issues and possible solutions. In many cases, the drafting of these chapters, through cooperation and collaboration, epitomizes how shared governance can work – through hard work, clear lines of communication, and sustained contact.

The first chapter to focus on a transboundary conflict (chapter 7) tells the story of the ongoing treaty negotiations for the Columbia River. Authors Paisley and Shurts provide insights on the complexities of this process. They describe the original treaty as successful within the context of the original design, yet grossly antiquated and overly simplistic in the modern context. Revisiting these mechanisms provides unique opportunities to remedy exclusions in the original versions – for example, the lack of representation from Indigenous peoples impacted by the treaty's provisions. In addition, this process raises questions about stakeholder participation and process, issues that are central to most of the flashpoints chapters. In particular, Locke and McKinney use the case of the Flathead River Basin (chapter 10) to show the importance of involving all affected parties in negotiations as early as possible. The consequences of ignoring this "golden rule" are often high (as seen in the Devils Lake example in chapter 9).

In addition to antiquated governing mechanisms, many of the flashpoint issues are complicated by management challenges posed by degraded infra-structure. In chapter 8, for example, Bankes and Bourget describe how controversies surrounding the apportionment of the St. Mary and Milk Rivers are complicated by degraded infrastructure such as leaky pipes. As in chapter 7, Bankes and Bourget highlight the need to continue to address issues in relation to changing physical conditions and the need to treat historical legacies with a contemporary lens.

For the Flathead River Basin (chapter 10), the issues are complicated by stark differences in interpretations of land-use planning on either side of the border. In this chapter, Locke and McKinney describe the controversy of the Flathead River between British Columbia and Montana. At its heart, the conflict arises from diametrically opposed regimes for managing land and water in Canada and the United States. The Flathead River Basin is one of the most protected watersheds in the United States. Upstream in Canada, however, the river has been zoned for resource development. These fundamentally incompatible poli-cies have created conflict between British Columbia and Montana – a conflict that has intensified from a local and regional issue to one with an international profile.

Perhaps the most controversial of the flashpoints – the Devils Lake Diversion and Red River Valley Water Supply (chapter 9) – grapples with myriad issues, including flooding, unauthorized inter-basin transfer projects, and unilateral action at a state and provincial level (North Dakota and Manitoba). In this case, failure to activate the established Canada-US dispute-resolution process has led to increased mistrust and heated rhetoric on both sides of the international border. The authors suggest steps to resolve the apparent impasse and under-score the importance of ongoing communication between actors involved in the governance of shared waters. Similar to the Flathead controversy, asym-metrical governance mechanisms and a breakdown in communication have exacerbated this conflict.

Moving away from conflict and controversy, the final chapter in Part II de-scribes how the Great Lakes can serve as a model for transboundary coopera-tion. Authors Linton and Hall illustrate in chapter 11 how changing physical, economic, and political environments can foster changes in institutional mech-anisms and governance structures by rendering challenges into opportunities for cooperation and collaboration. They argue: "the constant emergence of new problems and issues is balanced by a history of commitment to overcoming such challenges together." This penultimate chapter provides a positive example for other neighbours to follow, showing that although flashpoints may be inevitable, there are ways to prepare for these issues of shared concern.

The chapters in Part II are designed to highlight a number of key points. First, they show that both Canada and the United States play the role of "upstream aggressor" and "downstream aggrieved" in different situations; the Flathead River Basin (chapter 10) and Devils Lake (chapter 9) cases are perhaps the most controversial, yet one river flows north and the other, south. Second, the chapters showcase the diversity of issues that render transboundary water governance so challenging: irrigation and drainage, invasive species, mining and protected parks, and fisheries issues have all sparked transboundary water disputes. Finally, the chapters demonstrate a remarkable resolve on the part of key actors – sometimes policy-makers, sometimes organizations, and sometimes a single dedicated individual – to innovate and problem-solve at the border. Particularly noteworthy is that, in many cases, solutions have taken the form of working around – rather than through – the conventional institutions for transboundary water governance.

Conclusions

Since the 1990s, water governance has undergone a significant shift both *between* and *within* Canada and the United States. As detailed in chapter 3, new actors (such as citizens' advisory groups) are engaged in water governance at new scales (such as the watershed) in both countries (Cohen 2012; Norman and Bakker 2009; Sabatier et al. 2005). Within Canada, federal and provincial legislation imposes bans on inter-basin transfers; the federal government amended the Boundary Waters Treaty Act, banning the abstraction of bulk water from boundary waters that fall under the treaty.[8] Within the United States, although subnational actors are increasingly engaged in the governance of water resources through watershed boards and citizens' groups, and federal agencies are working closely with state agencies, municipalities, and non-state actors in water-related issues, the United States still faces many hurdles to achieving effective water governance. In particular, "the fragmented nature of US policies on water, a systematic tilt toward framing water as a market commodity rather than as a human right, the late arrival of US nongovernmental organizations to several important domains of global water politics, and some notable gaps between US policies" (Conca 2008, 215) all serve as barriers to effective water policy and governance in the United States. Fragmentation of institutional responsibilities is at the heart of this issue, particularly regarding how to make and pass legislation concerning both surface water and groundwater policy. Deason, Schad, and Sherk (2001, 175) aptly note that "With regard to institutional considerations in water resources policy development, an apparent oxymoron is true: there are both too many and too few actors."

Simultaneously, bilateral changes have taken place: notably, the North American Free Trade Agreement and the associated creation of the Commission for Environmental Cooperation (CEC).[9] New issues have also come to the fore, such as new types of pollutants and invasive species. In addition to these issues, the governance of transboundary waters between Canada and the United States is also complicated by broader trends in the relationship between the two countries, particularly with respect to environmental governance (Gattinger and Hale 2010).

The governance of shared waters is thus taking places in a dynamic, rapidly changing context. Within this broader context, this book brings together the world's foremost experts on Canada-US transboundary water issues to detail the changing fabric of the Canada-US relationship, analyse key freshwater flashpoints along the border, and assess the role of new and innovative transboundary water governance mechanisms. Our conceptual goal is to contribute to ongoing academic debates on environmental governance (e.g., the utility of watersheds as an organizing principle for governance) and on boundaries/borders. Our pragmatic goal is to explore how Canada and the United States might move forward with effective transboundary water governance in challenging and rapidly changing economic, socio-cultural, and biophysical environments.

NOTES

1 We define "transboundary" as crossing a state, provincial, or national border. We define "international" as being of, or relating to, two or more nations.

2 Since AD 805, the number has increased to 3,600 water-related treaties.

3 The only documented case of inter-state armed conflict over water took place between 2500 and 2350 BC between the Sumerian states of Lagash and Umma, although water has played a strategic role in numerous conflicts in history (Rasler, Karen A., and W.R. Thompson. 2006. Contested Territory, Strategic Rivalries, and Conflict Escalation. International Studies Quarterly 50 (1):145–168).

4 Bastmeijer, Bastmeijer, and Koivurova (2008); Conca (2006); Gleick (2000); Mitchell (1990); McGinnis (1999); Varady, Meehan, and McGovem (2009).

5 For example, Blomquist and Schlager (2005); Brun and Lasserre (2006); Cohen and Davidson (2011); Jarvis et al. (2005); Cumming, Cumming, and Redman (2006); Griffin (1999); Norman and Bakker (2009); Norman (2009).

6 For example, Budds (2009); Linton (2008, 2010); Swyngedouw (2006, 2007).

7 The authors – many of whom had never previously met – presented the first drafts of their chapters at a workshop held in Vancouver, British Columbia, Canada. The fruitful dialogue that resulted helped to identify and reinforce the core themes present throughout this volume. The workshop, which was funded by the Walter & Duncan Gordon Foundation, included commentaries from peers and invited officials that proved invaluable for this volume.

8 Bill C-15 is An Act to Amend the International Boundary Treaty Ac, S.C. 2001, 40. Bill C-15 specifically (a) "prohibits the bulk removal of boundary waters from the water basins in which they are located; b) requires persons to obtain licenses from the Minister of Foreign Affairs for water-related projects in boundary or trans-boundary waters that would affect the natural level or flow of waters on the United States side of the border; and c) provides clear sanctions and penalties for violation." This followed an IJC study calling for a moratorium on bulk water exports (International Joint Commission 2000), which stated that Canada and the US "should not permit any new proposal for removal of water from the Great Lakes Basin to proceed unless the proponent can demonstrate that the removal not endanger the integrity of the ecosystem."

9 Although the agreement primarily deals with trade disputes, a provision – Article 2004 – also provides for disputes dealing with environmental issues. The provision aimed to ensure that each member country knew about, and agreed to, the potential environmental hazards associated with trade issues. Specifically, the provisions allowed for the creation of panels of environmental experts, the submission of scientific advice, and the inclusion of public commentary.

REFERENCES

Bakker, Karen, ed. 2007. *Eau Canada: The Future of Canada's Water*. Vancouver: University of British Columbia.

Barlow, M. 2007. *Blue Covenant: The Global Water Crisis and the Coming Battle for the Right to Water*. Toronto, Ontario: McClelland & Stewart Ltd.

Bastmeijer, C.J., K. Bastmeijer, and T. Koivurova, eds. 2008. *Theory and Practice of Transboundary Environmental Impact Assessment*. Danvers, MA: Martinus Nijhoff Publishers.

Blomquist, W., and E. Schlager. 2005. "Political Pitfalls of Integrated Watershed Management." *Society & Natural Resources* 18 (2): 101–17. http://dx.doi.org/10.1080/08941920590894435.

Brun, A., and F. Lasserre. 2006. *Politiques de l'Eau: Grandes Principes et réalités locales*. Québec: Presses de l'Université du Québec.

Budds, J. 2009. "Contested H20: Science, Policy and Politics in Water Resources Management in Chile." *Geoforum* 40 (3): 418–30. http://dx.doi.org/10.1016/j.geoforum.2008.12.008.

Budds, J., and L. Hinojosa. 2012. "Restructuring and Rescaling Water Governance in Mining Contexts: The Co-Production of Waterscapes in Peru." *Water Alternatives* 5 (1): 119–37.

Cohen, A. 2012. "Watersheds as Boundary Objects: Scale at the Intersection of Competing Ideologies." *Environment & Planning A* 44 (9): 2207–24.

Cohen, A., and S. Davidson. 2011. "The Watershed Approach: Challenges, Antecedents, and the Transition from Technical Tool to Governance Unit." *Water Alternatives* 4 (1): 521–34.

Conca, Ken. 2006. *Governing Water: Contentious Transnational Politics and Global Institution Building.* Cambridge, MA: MIT Press.

Conca, Ken. 2008. "The United States and International Water Policy." *Journal of Environment & Development* 17 (3): 215–37. http://dx.doi.org/10.1177/1070496508319862.

Cook, Christina. 2011. *Putting the Pieces Together: Tracing Fragmentation in Ontario Water Governance.* PhD dissertation, Institute for Resources Environment, and Sustainability, University of British Columbia, Vancouver.

Cumming, G.S, D.H. Cumming, and C.L. Redman. 2006. "Scale Mismatches in Social-ecological Systems: Causes, Consequences, and Solutions." *Ecology and Society* 11 (1): Article 14.

Deason, Jonathan P., Theodore M. Schad, and George William Sherk. 2001. "Water Policy in the United States: A Perspective." *Water Policy* 3 (3): 175–92. http://dx.doi.org/10.1016/S1366-7017(01)00011-3.

European Union. 2011. "Water" website, http://ec.europa.eu/environment/water/index_en.htm. Accessed April 2011.

Forest, P. 2011. "Transferring Bulk Water between Canada and the United States: More Than a Century of Transboundary Inter-local Water Supplies." *Geoforum* 43 (1): 14–24. 10.1016/j.geoforum.2011.07.001.

Gattinger, Monica, and Geoffrey Hale. 2010. *Borders and Bridges: Canada's Policy Relations in North America.* Toronto: Oxford University Press.

Gleick, P.H. 2000. "The Changing Water Paradigm – A Look at Twenty-first Century Water Resources Development." *Water International* 25 (1): 127–38. http://dx.doi.org/10.1080/02508060008686804.

Gleick, P.H. 2007. *The World's Water: The Biennial Report on Freshwater Resources.* Washington, D.C.: Island Press.

Gooch, G.D., and A. Rieu-Clarke. 2010. *The Science-Policy-Stakeholder Interface and Transboundary Water Regimes. Science, Policy and Stakeholders in Water Management: An Integrated Approach to River Basin Management,* ed. G. D. Gooch and P. Stalnacke. London, UK: Earthscan.

Griffin, C.B. 1999. "Watershed Councils: An Emerging Form of Public Participation in Natural Resource Management." *Journal of the American Water Resources Association* 35 (3): 505–18. http://dx.doi.org/10.1111/j.1752-1688.1999.tb03607.x.

Heinmiller, B.T. 2008. "The Boundary Water Treaty and Canada – U.S. Relations in Abundance and Scarcity." *Wayne Law Review* 54 (4): 1499–524.

Hele, K.S. 2008. *Lines Drawn upon the Water: First Nations and the Great Lakes Borders and Borderlands.* Waterloo, Ontario: Wilfrid Laurier University Press.

Hill, C., K. Furlong, K. Bakker, and A. Cohen. 2008. "Harmonization versus Subsidiarity in Water Governance: A Review of Water Governance and Legislation in the Canadian Provinces and Territories." *Canadian Water Resources Journal* 33 (4): 315–32. http://dx.doi.org/10.4296/cwrj3304315.

International Joint Commission. 2000. "Protection of the Waters of the Great Lakes: Review of the Recommendations in the February 200 Report", Ottawa, Washington, DC. International Joint Commission. http://www.ijc.org/files/publications/ID1560.pdf

Jarvis, T., M. Giordano, S. Puri, K. Matsumoto, and A. Wolf. 2005. "International Borders, Ground Water Flow, and Hydroschizophrenia." *Ground Water* 43 (5): 764–70. http://dx.doi.org/10.1111/j.1745-6584.2005.00069.x. Medline:16149973Linton, J. 2008. "Is the Hydrologic Cycle Sustainable? A Historical-Geographical Critique of a Modern Concept." *Annals of the Association of American Geographers. Association of American Geographers* 98 (3): 630–49. http://dx.doi.org/10.1080/00045600802046619.

Linton, J. 2010. *What is Water? The History of Modern Abstractions.* Vancouver: University of British Columbia Press.

McGinnis, M.V. 1999. "Making the Watershed Connection." *Policy Studies Journal: The Journal of the Policy Studies Organization* 27 (3): 497–501. http://dx.doi.org/10.1111/j.1541-0072.1999.tb01982.x.

Mitchell, B. 1990. *Integrated Water Management: International Experience and Perspectives.* London: Bellhaven Press.

Norman, Emma S. 2009. *Navigating Bordered Geographies: Water Governance along the Canada-U.S. Border.* PhD dissertation, Department of Geography, University of British Columbia, Vancouver.

Norman, Emma S. 2012. "Cultural Politics and Transboundary Resource Governance in the Salish Sea." *Water Alternatives* 5 (1): 138–60.

Norman, Emma S. 2013. "Who's Counting? Spatial Politics, Ecocolonisatin, and the Politics of Calculation in Boundary Bay." *AREA* (Royal Geographic Society) 45 (2): 179–187. http://dx.doi.org/10.1111/area.12000.

Norman, Emma S., and Karen Bakker. 2009. "Transgressing Scales: Water Governance across the Canada-US Borderland." *Annals of the Association of American Geographers* 99 (1): 99–117. http://dx.doi.org/10.1080/00045600802317218.

Nowlan, L., and Karen Bakker. 2010. *Practising Shared Water Governance in Canada: A Primer.* Vancouver: Program on Water Governance.

Pentland, Ralph, and Adele Hurley. 2007. "Thirsty Neighbors: A Century of Canada-U.S. Transboundary Water Governance." In *Eau Canada: The Future of Canada's Water*, ed. K. Bakker, 163–84. Vancouver: University of British Columbia.

Phare, Merrell-Ann. 2009. *Denying the Source: The Crisis of First Nations Water Rights.* Calgary: Rocky ountain Books.

Reed, Maureen G. and Shannon Bruyneel. 2010. Rescaling environmental governance, rethinking the state: A three-dimensional review. 34 (5): 646–653.

Sabatier, Paul, Will Focht, Mark Lubell, Zev Trachtenberg, Arnold Vedlitz, and Marty Matlock. 2005. *Swimming Upstream: Collaborative Approaches to Watershed Management (American and Comparative Environmental Policy).* Cambridg, MA: MIT Press.

Simpson, A. 2000. "Paths Toward a Mohawk Nation: Narratives of Citizenship and Nationhood in Kahnawake." In *Political Theory and the Rights of Indigenous Peoples,* ed. D. Ivison, P. Patton, and W. Sanders. Cambridge: Cambridge University Press.

Simpson, A. 2007. "On Ethnographic Refusal: Indigeneity, 'Voice' and Colonial Citizenship." *Junctures: The Journal for Thematic Dialogue* 9.

Swyngedouw, Erik. 2006. "Circulations and Metabolisms: (Hybrid) Natures and (Cyborg) Cities." *Science as Culture* 15 (2): 105–21. http://dx.doi. org/10.1080/09505430600707970.

Swyngedouw, Erik. 2007. "Technonatural Revolutions: The Scalar Politics of Franco's Hydro-social Dream for Spain, 1939–1975." *Transactions of the Institute of British Geographers* 32 (1): 9–28. http://dx.doi.org/10.1111/j.1475-5661.2007.00233.x.

Thom, Brian. 2010. "The Anathema of Aggregation: Towards 21st-century Self-government in the Coast Salish World." *Anthropoligica* 52:33–48.

United Nations. 2011. "UN Water" website, http://www.unwater.org/. Accessed April 2011.

UN World Water Assessment (UNWWAP). 2012. *Managing Water under Uncertainty and Risk.* Paris, France: United Nations Educational, Scientific and Cultural Organization.

Varady, R.G., K. Meehan, and E. McGovern. 2009. "Charting the Emergence of Global Water Initiatives in World Water Governance." *Physics and Chemistry of the Earth* 34 (3): 150–5. http://dx.doi.org/10.1016/j.pce.2008.06.004.

Vörösmarty, C.J., P. Green, J. Salisbury, and R.B. Lammers. 14 Jul, 2000. "Global Water Resources: Vulnerability from Climate Change and Population Growth." *Science* 289 (5477): 284–8. http://dx.doi.org/10.1126/science.289.5477.284. Medline:10894773

Vörösmarty, C.J., P.B. McIntyre, M.O. Gessner, D. Dudgeon, A. Prusevich, P. Green, S. Glidden, S.E. Bunn, C.A. Sullivan, C.R. Liermann, et al. 2010. "Global Threats to Human Water Security and River Biodiversity." *Nature* 467 (7315): 555–61. http://dx.doi.org/10.1038/nature09440. Medline:20882010

Wolf, Aaron T. 2001. "Water and Human Security." *Journal of Contemporary Water Research and Education* 118:29–37.

Wouters, P., and D. Ziganshina. 2011. "Tackling the Global Water Crisis: Unlocking International Law as Fundamental to the Peaceful Management of the World's Shared Transboundary Waters – Introducing the H 2 O Paradigm." In *Water Resources Planning and Management: Challenges and Solutions,* ed. Quentin Grafton and Karen Hussy, 175–229. Cambridge: Cambridge University Press.

PART ONE

Issues, Approaches, and Challenges

2 Indigenous Peoples and Water: Governing across Borders

MERRELL-ANN S. PHARE

A border is a boundary. In its simplest and most basic sense, a border is a severing of space created by a physical, mental, legal, or social separation. Sometimes these boundaries are subtle, like the lines on a page that tell a child where or where not to place a crayon. At other times, they are walls made of wood and concrete that, when placed in front of us, can hide us from those on the other side. Looking around us, borders seem ubiquitous: the edge of my hand as it touches this keyboard, the edge of a mountain against the cloudless blue sky, or the edge of a river flowing along its bank.

These borders are false. While it is said that nature abhors a vacuum, it is even truer that nature forbids borders. Quantum physics has shown no real boundary exists between the electrons that make up the cells in my hand as it touches those of my computer keyboard. A mountain, through each of the minuscule pebbles that it sheds, becomes something else: oil, dust, or a riverbank. A river is merely a few adventurous molecules that surfaced from a complex network of underground wells and veins of water into the vast world above them. These molecules then move on to some other place, evaporate, fall as rain, and then are potentially taken up by a plant, continuing through ever-moving, perpetual cycles of transformation.

Indigenous peoples worldwide have long spoken of this deep connection between all things (Lavalley 2006, 63), with water being the archetypal example of borderlessness. Water is the great unifier. To traditional Indigenous peoples, the Earth is mother and waterways are her veins, capillaries, and arteries, through which water flows and nourishes everything: "Water is a relation, and it connects us to all other living things in the 'web' of life" (Chiefs of Ontario 2006, 10). In this worldview, some borders – separations between us and them, now and then, this and that – can be helpful in accomplishing a goal, such as limiting human behaviour or protecting a sacred place. However, these human-created lines are not reality. And, all too often, these separation systems can fail us when we fail to see the truth they obscure. This truth, as the natural world shows us, is that everything is everything.

Since colonization, Indigenous peoples have struggled to maintain their sense of community and collective identity in the face of borders imposed by Western laws and culture. Indigenous people have experienced numerous and continued intrusions of "the border" into their lives in the form of restriction of movement, thought, and entitlement. This book is about these borders in the context of across the Canada – US border. However, I submit that for most Indigenous peoples, the borders drawn around their nations by the creation of "reserves" in Canada and "reservations" in the United States, have been more significant, particularly in terms of their continued ability to be in a relationship with water and manage themselves, their communities, and water in furtherance of that relationship.[1]

Many Indigenous people are "intensifying the development of strategies concerning the border" (Miller 1996–7, 64), trying to erase, cross, or work around the lines we have created. They are developing strategies to the benefit of water, their people, and nature, and as a way to achieve a rebalancing of social justice in their favour. However, their stories go largely untold, particularly in the academic literature (Miller 1996–7, 64). This chapter attempts to share the stories of three such people and the strategies they are using to dismantle the borders that tend to deny social justice to their people.

Case Study 1: Unama'ki Institute of Natural Resources

Shelley Denny lives and works on a border between two knowledge systems, trying to find ways for these systems to communicate with each other to save a lake. These two knowledge systems are the Western sciences and the traditional knowledge of Indigenous peoples. Denny works as a fisheries biologist and is now research director of the Unama'ki Institute of Natural Resources (the UINR) in the area where she spent her childhood: the Bras d'Or Lake region of the island of Cape Breton, Nova Scotia. I first met Denny in 2002 at a meeting on environmental management that we both attended in Winnipeg, Manitoba. Although only in her early thirties and recently out of graduate school, she was already playing a lead role in watershed management in Nova Scotia.

Denny is passionate about water, people, and the borders between the two. However, she does not describe her passion this way, as being focused on borders; instead, she talks about her work in places like wetlands, shorelines, and lakebeds, where the needs of water and of her people *overlap* (often to the detriment of the lake). Since 2005, through her work at the UINR, Denny has worked closely with the five First Nation communities around the Bras d'Or Lake that created the UINR: the Eskasoni, Membertou, Potlotek, Wagmatcook, and Waycobah Nations. All are Mi'kmaq and Denny herself is a Mi'kmaq woman. She grew up living with her grandmother, playing in, and learning from, the waters in this area. She recalls summers spent bent over, hands on her knees, staring at the many skittering creatures that live in the narrow strip of habitat along the banks and in the stream near her grandmother's house: "I was either swimming in it, skating on it, or learning from it. Water is me."

To this day, she is still amazed at the beauty of the lake, and as a scientist, at the diversity of species living in the few metres of marsh along the water's edge. The Bras d'Or Lake has been famously referred to by Canadian author Silver Donald Cameron as "a basin ringed by indigo hills laced with marble. Islands within a sea inside an island." (See Figure 2.1.) It is a large body of water, measuring almost 100 kilometres long and 50 kilometres wide, dominating the Cape Breton Island landscape. It has two natural eastern connections to the Atlantic Ocean through the Cabot Strait and one southern channel constructed in 1869 at St. Peters. The lake's top layer is freshwater fed by six major rivers that empty into the lake, and the bottom layer of water is comprised of the heavier saltwater from the Atlantic Ocean. There are also variable and unique brackish combinations of the two throughout the lake.

Figure 2.1 The Bras d'Or Lake Drainage Basin within First Nations' Traditional Territories.

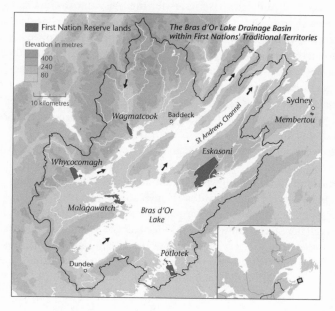

The species here also blur boundaries, particularly the distinction between marine and freshwater species. One local creature is the black-spotted stickleback, a lemon-yellow fish sporting black spots and two spines along its back; the males build their nests over sandy bottoms using small leaves and wood for construction materials. The lakes are also home to the American oyster, which is currently under great threat from an invasive US species: the deadly MSX parasite. The sand shrimp, a largely colourless crustacean with many tiny, star-shaped spots that make it almost indistinguishable from the sand below, also lives here.

According to Denny, the Bras d'Or is also home to the American eel, *Anguilla rostrata*. This eel is one of the most important staple food sources for First Nations around the lake. Denny remembers her grandmother's face upon receiving a gift of eel from other community members. "Better than Christmas," her grandmother would say. These eels, whose colours change from transparent to yellow to silver as the eels mature, are long-distance travellers that spend most of their lives in the surficial freshwater layer of the Bras d'Or, living in the eelgrass beds that are found along the lake's edge.

Denny has discovered that the main threats to the health of the Bras d'Or Lake and all its species are the activities of people who use and, in many cases, depend upon the lake. Sewage from local homes, as well as fertilizers and others chemicals that run off from nearby farms and lawns pollute the waters of the Bras d'Or. Fecal coliform bacteria contamination from sewage run-off has severely impacted population numbers for the normally hardy American oyster. Removal of trees and shrubs from along the water's edge has destroyed the ability of these plants to take up nutrients from the soil and purify water as it travels down to the lake. The eel and its habitat, eelgrass, have been damaged by tree removal, which causes increased levels of sediment to flow down into the lake from deforested riverbanks. As a 2009 project of the UINR that mapped traditional knowledge of eelgrass sites reported, "'No eel grass, no eels.' That's the simple truth as told by Mi'kmaq knowledge holders around the Bras d'Or Lakes" (Denny 2009, 3).

Denny has been striving to change the behaviour of people who live and work around the Bras d'Or, the approach of scientists studying the lake, and the attitudes of those who govern lake management. She is starting with her own people by trying to generate awareness about the impacts of simple, individual actions, primarily by focusing on the species that they value the most. Denny is relying on the knowledge of the five local Mi'kmaq nations to remind those who live near or care about the lake, First Nations and non-First Nations alike, of the connections between the lake and the people. Bridging even the smallest gap in understanding helps to improve the ways that humans manage themselves around the lake.

When Denny explains what she does, she talks about building relationships with the lake, its spirit, and its species. Denny speaks most often about one species in particular: eels. Traditional knowledge – about how the eels behave during thunderstorms, how they layer themselves in the lake bottom when they overwinter, and how the colour of an eel's den indicates whether it is being actively used or not – contributes to a more detailed understanding of the lake. Denny is considering creating an eel-classification system, but not based on the usual criteria, such as habitat or species differences. Instead, her system would be based on traditional uses of the eels by the Mi´kmaq; the classification system would be defined by factors such as age, location, and time of year harvested. To Denny, these criteria are aimed at ensuring the healthy continuation of the eel population. "Scientists don't come out here often enough to really know the eel. My people live here, they have watched eels forever," she says, as she explains the eel knowledge she is using to build a best practices manual for lake management. Western science has helped identify the source of many problems affecting the lake, but Denny is relying on the knowledge of her people to find the solutions.

Since Denny relies on both Western science and traditional knowledge to formulate better management and health-restoration strategies for the Bras d´Or Lake, she is employing a practice called "two-eyed seeing." Mi´kmaq Elder Albert Marshall teaches that the solution to many of our environmental challenges can be found if we practice this form of understanding the world, with one eye seeing the best knowledge and values of the Western world, and the other seeing the best knowledge and values of First Nations peoples.

Two-eyed seeing adamantly, respectfully, and passionately asks that we bring together our different ways of knowing to motivate people, Aboriginal and non-Aboriginal alike, to use all our understandings so we can leave the world a better place and not comprise the opportunities for our youth (in the sense of the Seven Generations) through our own inactions (Marshall 2011).

This is the "in-between space," the place of a new way to the future. Denny, a translator of worldviews, lives in this space, refusing to accept the limitations and barriers that exist in science.

Border as Process

A border is a construct of our own invention that we use to create structure and order. While borders are sometimes physical, words and traditions can also create borders. We often refer to this second type of border as "the rules," but sometimes, they are just "the way we do things around here." We hold these borders close to us, and from deep within our psyches, they limit our behaviour, help maintain peace, and make us conform. We give them immense status in our societies by codifying them as laws and regulations.

Many of these laws, particularly the ones related to water, have excluded Indigenous peoples' values, their influence on decision-making, and the unique impacts they experience. Western water laws and policies have created a separation between Indigenous peoples and the rest of society, so Indigenous ideas about interconnection, responsibility, and reciprocity have not infiltrated Western ways of thinking about water.[2] Many examples demonstrate how the infiltration of these ideas would be helpful. For instance, people in the water policy arena typically talk about "managing water" instead of managing ourselves in relation to water. The choice of language is not an issue of semantics. It demonstrates the difference between a commitment to demand management and the construction of another new dam. A second example: upon recognizing a problem in our current system, we usually solve it by identifying the appropriate jurisdiction, liability, and responsibility, and then (if we are feeling particularly cooperative) creating some agreement to better meet human water needs. Transboundary water agreements are a fine example of this practice, and they often result, at least for a while, in better water allocation, the protection of water sources, and increased communication across jurisdictional boundaries. What they are largely doing, however, is working (albeit very successfully in some circumstances) around the limitations imposed by the challenging and unnatural jurisdictional boundaries we first created ourselves.

Indigenous peoples are trying to break through these borders by involving their people and governments in processes of their own making that are intimately based on their cultures and values, and that address human and ecosystem needs for water.

Case Study 2: Collaborative Environmental Planning Initiative (CEPI) and The Spirit of the Lake Sings

I was conceived by the Creator on both sides of an ancient ocean, then forged in the tropics within the middle of a supercontinent; all the while being weathered by the Elements of fire, wind, ice, and water over some 600 million years. The final form of my embryo was developed while entombed in ice some 22,000 years ago. I was birthed 11,000 years later as the glaciers receded and their melt waters poured life into my streams and underground waters. During the first 5,000 years of my childhood I was in the form of fresh water lakes, rushing headlong to the distant sea through a turbulent mountain stream confined by the narrow, deep gorge of the Great Bras D'Or Channel .... Some 6,000 years ago the sea rose to meet me, lapping higher and higher on my shores, allowing my evolution into the estuary that bathes and blankets the parts of me you see today. (CEPI 2010, 6)

The Bras d'Or Lake has a number of connections to the sea, including a northern connection to the Atlantic Ocean through the Middle Shoal channel. International shipping companies make heavy use of this channel as a navigational entrance to the Bras d'Or. In 1996, a battle by the Mi'kmaq to protect the natural state of the channel launched a multigovernmental cooperative approach for protecting the Bras d'Or Lake; the approach is called the Collaborative Environmental Planning Initiative (CEPI).

Dan Christmas, a member of Membertou First Nation in New Brunswick and chairperson of CEPI's Management Committee, noted the importance of this event: "A shipping company wanted to get bigger loads on their ships. They were leaving Bras d'Or half full and they wanted to leave fully loaded and to do that, they had to dredge the bottom of the mouth of Bras d'Or. We felt very strongly as Mi'kmaq that this would negatively impact the migration of fish in and out of the lake, and we insisted that there had to be some kind of scientific work before the permits were issued." Christmas further explained that "the courts agreed with the Mi'kmaq, and after that legal decision the Mi'kmaq became a stronger force in protecting the lakes. The government realized that they had to work with the Mi'kmaq. That was the germ of what later became CEPI."

CEPI is a multigovernmental approach to addressing environmental concerns around the Bras d'Or Lake. The CEPI charter commits its signatories – the Mi'kmaq people, as well as the municipal, provincial, and federal agencies of the region – to bring their laws, expertise, and interests to the group to collaborate in building solutions for the Bras d'Or Lake. These important features – comprehensiveness and a focus on collaboration – have also created problems. Despite this initiative's critical purpose, it has been challenging to implement, largely because of the jurisdictional territoriality of many of the CEPI signatories. "In particular," Christmas explains, "there has been reluctance to acknowledge the roles of First Nations, who comprise only ten percent of the population around the Bras d'Or but who have constitutionally-protected rights to fish and harvest from its waters." After a decade of increasing challenges plaguing CEPI, the partnership was in danger of disintegrating. Thus, in 2008, the Mi'kmaq communities asked Christmas to review the CEPI process and find a way to improve its implementation. *The Spirit of the Lakes Speaks,* the final report of Christmas's review of the CEPI process, was the result.

Christmas was asked to play this pivotal role because he had worked diligently on these issues for some time, having participated in the first unsuccessful efforts to create a management plan for the lake almost 20 years ago. "It seemed then that the Bras d'Or Lake was broken up into many overlapping jurisdictional debates, and what was being lost in the middle of course was the lake itself, the ecosystem, and the watershed. It was almost jurisdictional paralysis. Many issues were happening in the lakes and we couldn't figure out who was responsible to do what. I remember at one point, DFO had licensed some draggers to go inside the lakes and drag the bottom and this was horrifying for all the Mi'kmaq elders, and because they knew some species were very sensitive and at that time, it seemed like we were totally powerless to stop that. That's when things changed for us."

In many ways, *The Spirit of the Lakes Speaks* is like Christmas himself: straightforward, patient, and judicious in approach. Through this experience, Christmas learned the importance of having a clear process with defined goals directed at the lake itself rather than at jurisdictional preservation. Involvement, commitment, and willingness to focus on the lake as the first priority were the necessary factors to decrease human impacts. This approach was a dramatic change from the previous one, which involved starting discussions based on the scope of the various jurisdictions at play. *The Spirit of the Lakes* sets out the process plan for management of the lake based on the four quadrants of the Elders' Medicine Wheel teachings: knowledge, action, spirituality, and feelings. To have a healthy lake, these four quadrants must govern decisions made about the Bras D'Or Lake, and, according to Christmas, in so doing, they will be "connected to the lake through the people." The plan does not set out what must be done; this work can only be fulfilled by the people who live along the lake and supported by the necessary information, resources, and institutions (such as the UINR). Instead, the framework guides their actions through the principles and structure of a Medicine Wheel – inspired planning cycle. "Once we just got over the jurisdictional reluctance and got to focus on the issues of the lakes and the species in the lakes, it was then that everyone started to put their boundaries or divisions or walls down, and say let's talk about the species, what's wrong with the oysters, what's wrong with the herring. Once the discussion became lake-based or species-based, it seemed that those divisions or those lines on the map, so to speak, began to disappear," Christmas explains.

Christmas believes that with this new process in place, the vision of CEPI will be achieved in his lifetime. "We are still 20 years away, I think," he says smiling. "But in the end I see a picture of a pristine lake that is not polluted. I see an environment that is productive, meaning that the species are healthy and thriving and I see people, not just Mi'kmaq, but people along the lakes, enjoying the beauty of the lakes, being able to swim, being able to fish, being able to skate on the lakes in the winter time, and ice fish. I see people building their homes and cottages, in a way that is environmentally friendly, with no sewage seeping into the lakes and we are not using fertilizers, there are no foresters cutting right to the shoreline, or along side brooks and streams. I see an environment that is being productive and that people can enjoy, just not physically but spiritually as well."

Borders as a Place of Reconciliation

While many may focus on the spaces that borders create on either side of themselves, arguably, the most important space they create is the thin line of the border itself. If one were to look very closely at this line, say through the lens of a powerful microscope, the edges of the border would likely be impossible to see.

What would be visible are the details of the line itself: the small gaps where the dark colour becomes lighter and more transparent, the areas where its texture and uniqueness become apparent. If it were possible to stand right on that line, the space around you would look like its own unique landscape for as far as you could see. Not unlike the world around you at this moment, you would likely be unable to point to the exact edge of the space you inhabit on the line. The horizon would be simply the last part of the line that is visible to your eye, an edge in relation to you, perhaps, but maybe not the actual edge. From that perspective, it would be clear that within the line resides a massive space where there is room to move, where the boundary is actually unknown. In this sense, a border is a space of possibility: a space of the future in a world that is different from, possibly even better than, the one we have currently created.

How do we achieve a better world where water and the people and ecosystems that depend upon it are not under threat? One helpful concept is found in an idea the Supreme Court of Canada (SCC) presented in relation to Aboriginal rights cases. The SCC has repeatedly stated that the primary purpose of consulting with Aboriginal groups is to achieve reconciliation between Aboriginal and non-Aboriginal peoples, given that "we are all here to stay."[3] While reconciliation is considered a goal specifically when evaluating cases about infringements on Aboriginal rights, those in the field of water governance should consider this sage concept. A person's place, and the choices he or she makes while there, potentially can: erase boundaries between the past and the present and future, recognize and restore the relationships between all parts of nature, and allow for the development and implementation of partnerships to preserve a thriving future for all that depend upon Canada's lands and waters. As such, each place and the choices made within them remove boundaries that act as barriers to better water-related governance. The world viewed from the place where each one of us stands can be a place of reconciliation with water.

Case Study 3: Government of Northwest Territories and the Northern Voices, Northern Waters Water Stewardship Strategy

As I was flying into Yellowknife in January 2011 to attend a meeting about water, it became clear to me why Government of Northwest Territories (the GNWT) Deputy Premier Michael Miltenberger knew that creating a water strategy for the Northwest Territories (the NWT) was one of the most important things he could do during his political tenure. From the air, the land below appeared to be as much lake and river as solid ground. This is topography is the most defining characteristic of the NWT, and it is a stunning sight: sinkholes, potholes, springs, rivers, and lakes dot the landscape for as far as one can see. Indeed, the NWT is home to Canada's largest watershed, the Mackenzie River Basin (see Figure 2.2), and Yellowknife is situated on the northern shores of Canada's deepest lake, the Great Slave Lake. The Mackenzie River Basin sports many other distinctions, including housing Canada's longest river, the Mackenzie River, at 4,241 km; largest drainage area at 1.8 million km2; and Canada's largest freshwater delta, which is the twelfth largest in the world and covers a surface area of 13,500 km2 (Government of Northwest Territories 2010, 7).

Figure 2.2 The Mackenzie River Basin.

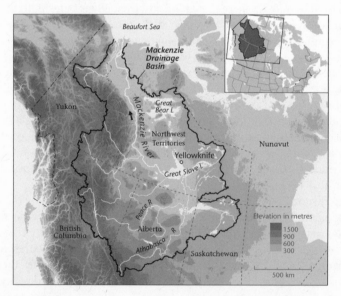

Miltenberger, who is also the Minister of Finance and Minister of Environment and Natural Resources, became connected to this land when he was just 11 years old and his family moved to Yellowknife from Eastern Canada. His first memory of the NWT is of "a lot of water, rocks, and rain," but he knew from very early on that it was here, in this landscape, where he would build his life. He now lives in Fort Smith, just steps away from nearby rapids of the Slave River. These same rapids have so far survived repeated threats in the form of unrelenting interest in the river's huge hydroelectric development potential. A long-time politician, Miltenberger is Métis, and, like other Indigenous leaders of the NWT and northern Alberta, has led increasingly loud calls for a consensus approach to protecting these waters. The now-yearly Keepers of the Waters conference was first convened in 2006 in Liidlii Kue, Denendeh, in the NWT, and was instrumental in beginning to create a unified voice. Then, in 2007, the GNWT declared its support for a motion Miltenberger presented that "all peoples have a fundamental human right to water that must be recognized nationally and internationally, including the development of appropriate institutional mechanisms to ensure that these rights are implemented." Other initiatives followed, including the Sahtu Water Gathering in Fort Good Hope (2008) and the National Summit on the Environment and Water (2008), hosted by the Dene Nation.

Why the mounting efforts? As Miltenberger explains, the waters of the NWT are under serious pressure from numerous potentially devastating threats:

Upstream development in the Mackenzie River Basin, including hydro, nuclear, and oil sands development have or may have an impact on our aquatic ecosystems. In northeastern British Columbia, the Bennett Dam has already altered river flows. BC Hydro has recently proposed a 900-megawatt project to develop Site C, the third dam on the Peace River, which may continue to modify flows. Downstream Aboriginal Governments have indicated concern over even slight additional alterations. In Alberta, TransAlta continues to gather information to develop a hydroelectric project for a 100-megawatt facility and a four-reactor 4,000 megawatt nuclear power plant on the Peace River. In the Great Slave sub-basin of the Mackenzie River Basin, a proposal for a Slave River Hydro Development near the NWT border is being investigated. Oil sands development continues in the Athabasca sub-basin of the Mackenzie River Basin.

Another concern is the massive impact of hydraulic fracturing for shale gas in upstream watersheds in British Columbia. This hazardous activity is occurring largely due to the lack of a coherent regulatory system for water withdrawals for this purpose (Parfitt 2010, 44). Residents of the NWT recognize that these various pressures, combined with the effects of climate change, could permanently alter the NWT, as well as everything and everyone who lives there and depends on the region's continued health. These concerns resulted in *Northern Voices, Northern Waters: The NWT Water Stewardship Strategy*, which, like its strongest proponent, is a prime example of vision and leadership surfacing when it is most needed.

Northern Voices, Northern Waters is based on the approach the UINR is experiencing success with at the other end of Canada. In the NWT, Western and traditional Indigenous values, knowledge, and science are combining in recognition of the interconnections that make life in the cultural, legal, and physical landscape possible. The strategy sets out an ecosystem-based approach within the basin, acknowledging that decisions concerning the use of waters within the basin potentially may affect many ecosystems. "These decisions cannot be made in isolation. Decisions and subsequent actions must be made after considering the entire watershed, its land and water, and all the values within it," Miltenberger explained when introducing *Northern Voices, Northern Waters* to the GNWT legislature on 20 May 2010. In addition, the strategy recognises the rights of Indigenous peoples in the NWT, with a focus on preserving cultural uses and relationships with water (Government of Northwest Territories 2010, 4).

The *Northern Voices, Northern Waters* strategy is a visionary political and policy statement, setting the critical foundation necessary to protect the people and ecosystems of the NWT; however, it is not without its challenges. Consensus support for the strategy is strong, but not infallible. Miltenberger explains that "the strategy was created because we all agreed water was more important than any of our other concerns. There are almost 45,000 of us here, and we can fight about almost anything. But water is different. We put those disagreements aside to get this done." Although the strategy currently has the support of the First Nations, Métis, and Inuvialuit of the NWT, the Akaitcho did not participate in its development. This group has not yet settled its land claim and has concerns about how the strategy could potentially affect its rights and interests. The Akaitcho chose instead to observe the process and have reserved the right to opt into the strategy after their nation's claim has been settled.

Another challenge is that the GNWT is undergoing significant changes as a government, including devolution: governmental authority over NWT lands and waters currently held by Canada will soon be transferred to the GNWT. Some worry that this process might threaten the stability of settled land claims and prevent the full implementation of the associated commitments. Of particular concern are the processes and structures, such as land and water boards, created under land-claim agreements to manage, develop, and protect lands and waters. It is not only politically and morally untenable, but also legally untenable, to implement devolution at the expense of current or future land-claims agreements. "These initiatives must strengthen Indigenous voices and rights if *Northern Voices, Northern Waters* is to be implemented successfully," Miltenberger states.

Chris Wood, an environmental journalist who recently wrote extensively on current challenges in the Mackenzie River Basin, describes the many existing threats to consensus:

> ... a territorial legislature struggling for traction in a constitutional vacuum; the Northwest Territories' fitful embrace of a new relationship with its indigenous Dene (Sahtu, Tlicho, Gwich´in, Akaitcho, and Dehcho), Inuvialuit, and Métis people; those same First Nations' increasingly muscular assertion of rights granted under treaties both historic and modern; and doubts about the ability of science to measure, let alone limit, industry's imprint on the wild. (Wood 2010, 28)

The strategy is also raising some eyebrows from upstream development interests in British Columbia and Alberta. Miltenberger welcomes the attention, fully aware that *Northern Voices, Northern Waters* takes a perspective that may not be shared by those to the south, maintaining that upstream interests have a legal and ethical obligation to protect the rights and needs of the NWT people and ecosystems. "The Mackenzie River Basin Transboundary Waters Master Agreement was signed by the GNWT, the upstream Governments of Saskatchewan, Alberta, and British Columbia, as well as Yukon, and Canada. This agreement was supposed to maintain the ecological integrity of the Mackenzie Basin ecosystem by ensuring that no water use unreasonably harm the ecological integrity in any other jurisdiction. It came into effect almost fifteen years ago, and yet there has never been even one meeting of the parties to this agreement. It has never been implemented, but development that has negatively affected the waters that come into the NWT has proceeded anyway." According to Miltenberger, a primary goal of the *Northern Voices, Northern Waters* strategy is to challenge any notion that the political and legal borders in northern British Columbia and Alberta absolve those governments of responsibility for degraded waters flowing into the NWT.

Clearly, implementing the strategy may pose difficulties, but the leadership shown by the GNWT, as well as the Indigenous peoples of those lands and waters, is noteworthy. The creators of *Northern Voices, Northern Waters* have approached water resource protection with the acknowledgment that the needs of water must come before the needs of humans, and that water has stewardship responsibilities to people and ecosystems. As pointed out by Robert Sandford, chair of the United Nation's Water for Life Decade in Canada, the NWT has acted in a unique way to protect its waters before a crisis could develop (Robert Sandford, personal communication, 7 January 2011, Yellowknife, NWT). Like Denny and the UINR, as well as Christmas and the CEPI, the GNWT and its partners have approached water protection, use, and management strategies from the perspective of the basin and its needs. These three groups have also refused to become entrenched in the limitations of their respective jurisdictional boundaries or unresolved issues. Miltenberger, in partnership with the Indigenous peoples of the NWT, is trying to move toward reconciliation between people and ecosystems, and dissolution of the artificial separation humans have created between them.

Conclusion: The Path Forward

Native American legal scholar and philosopher Vine Deloria Jr. described the difference between Indigenous and Western philosophies as of one of space and time:

> Native Americans hold their lands – places – as having the highest possible meaning, and all their statements are made with this reference point in mind. Immigrants review the movement of their ancestors across the continent as a steady progression of basically good events and experiences, thereby placing history – time – in the best possible light. When one group is concerned with the philosophical problem of space and the other with the philosophical problem of time, then the statements of either group do not make much sense when transferred from one context to the other. (Deloria 2003, 61)

In terms of people and their use of water, there is a third element of difficulty: a time-space overlap created by the unique characteristics of water, which moves from location to location over time along its cycle. This makes it challenging for people of different jurisdictions to understand one another's perspectives and relationships to land, water, and place, including the elements of responsibility we all have to water.

Moreover, as John Borrows explains, our contemporary environmental management approaches "were developed within a cultural logic that erased prior Indigenous presence and ecological relationships." Thus, while "Indigenous inclusion and involvement in existing institutions facilitates sustainability by suggesting important reconnections of biological relationships within ecosystems" (Borrows 2002, 34), even a fulsome inclusion of Indigenous perspectives in existing institutions may no longer be enough to address the water challenges we face. Indigenous peoples, and by extension their laws and processes, can be woven into the fabric of our cultural and legal landscape, but doing so "brings into sharper relief the socially constructed notions of space currently passing as neutral facts" in those landscapes (Borrows 2002, 35). The most significant "neutral fact" is the short-term and narrowly local basis of current institutional decision-making, which avoids serious consideration of upstream, downstream, or beside-the-stream concerns. What is most required at this time is serious reconsideration of our choices and actions regarding water.

On the last page of his book, *God is Red*, Vine Deloria Jr. asks a poignant question: "The future of humankind lies waiting for those who will come to understand their lives and take up their responsibilities to all living things. Who will listen to the trees, the animals and birds, and the places of the voices of the land?" (Deloria 2003, 296). The case studies in this chapter indicate that new relationships, approaches, and ways of thinking are needed to address the complexity of our challenges in protecting and sustainably using water across legal, cultural, and physical borders. Each study has shown that we can make several choices towards accomplishing this – from the local to the national, the scientific to the political, and the personal to the institutional.

First, we need leaders. We need to expand our conception of what leadership means and recognize Indigenous-led initiatives. While leadership comes in many forms and operates at many levels, Evelyn Pinkerton's (1991) research, in the context of Indigenous communities and fisheries-habitat management, indicates that supporting the efforts of leadership at the community level can have the greatest impact 1,332). Indeed, she has found that this approach is far more successful than government-directed or regulatory approaches alone. These initiatives allow for the creation of "issue networks" and multi-level, cross-border coalitions that encourage debate and discussion, producing co-governance models with a greater chance of success. As Burleson (2011) notes, "Political conversation builds skills unattainable by merely instilling political knowledge" (247). Recognition of community-based leadership initiatives creates the foundation for later adoption of regulatory measures and governmental policy, when and where appropriate. Indigenous peoples generate and share values based on their experience that they have real control over the situation and that they can protect their watersheds using strategies viewed as "our way," rather than imposed by outsiders. However, as the UINR and CEPI have demonstrated, these initiatives also exhibit strength in cross-border partnerships of people, institutions, and knowledge. As the GNWT implements its *Northern Voices, Northern Waters* strategy, it will need to address the challenging task of articulating its issues in ways that identify with the public interest, particularly concerning the governments of British Columbia and Alberta.

Second, we must be willing to envision a world where water and humans are in a partnership of mutual responsibility that has an ethical foundation. At first, it may be difficult to articulate what this vision means exactly, given that many of us likely are far from engaging in this form of relationship with a natural resource. Calling for this shift in perspective is not a politically acceptable position to hold in many parts of Canada today, since it would result in different decisions being made about how we use water and share it with nature. This outcome is unavoidable; indeed, this shift is the goal. This change in perspective would create an opening in the false boundary we have fashioned between humans and nature, providing opportunities from which we all can benefit. To accomplish this shift, we may be required to decide that, like the residents of the NWT, "we care more about water than our rights" and other current water management institutions, legal and otherwise, that we hold dear. These changes are what we need. We also need to recognize the leadership of Indigenous peoples, acknowledging the existence, legitimacy, and value of their initiatives, as well as the associated cultural perspectives and foundations of these initiatives. Recognizing and learning to value Indigenous knowledge and experience would allow society to create different principles for managing ourselves and our relationships with water. This influence is where the leadership of Indigenous peoples may have the broadest effect, in terms of creating long-standing change that is necessary to ensure water security. As Pinkerton (1993) notes, these tactics "have shown promise in improving the management of fisheries, forests, wildlife, water, and other common pool resources in an ecologically and hence economically sustainable direction" (37).

Using these examples, we can derive that process matters, in all its iterative, shifting, and multi-party forms. We need a voice and we need it to be heard; processes are the places where the voices we have are shared and heard. In addition to creating and committing to processes that promote reconciliation and build equity, we will need to ask ourselves whether nonhuman others, such as the waters themselves, have a voice.

Third, we will likely need new institutions. Each case study demonstrated the need for the creation of new institutions, some that address new ways of thinking and using knowledge, and others that are legal, cultural, or physical. Each institution began with a new approach to addressing human and ecosystem needs for water, and each has a champion with the vision and willpower to create a sense of insistence in addressing those goals, even in the face of other competing issues. This goal means recognizing that ensuring enough water for all people and ecosystems is the human survival imperative. When contemplating how we can better govern ourselves so that water is plentiful for future generations of all species, two-eyed seeing holds great promise as a method for people and institutions to achieve better management across borders. The limits to our success through current governance structures exist, whether we see them or not, which is why two-eyed seeing mandates that we re-evaluate everything we accept as truth in both of our knowledge systems. Only then will we able to choose which decisions, behaviours, laws, and policies from each of these knowledge systems make the most sense in the current world we face, and the world we want for our future. Governing at watershed scale – such as the IJC's IWI (chapter 4) – is a modest start, but we still have a long way to go.

Ironically, we must challenge borders to create limits. We must be willing to challenge the rules of our knowledge systems, as well as the legal, governmental, and academic institutions that rely on them. We must acknowledge that many of these institutions cling to outmoded ways of thinking, which exclude wisdom and data about our inseparable connection to nature and the limited availability of the world for our sole use. Community-based knowledge can help close these information gaps (Pinkerton 2008, 161).

Fourth, "the peculiar genius of each continent – each river valley, the rugged mountains, the placid lakes – all call for relief from the constant burden of exploitation" (Deloria 2003, 296). If all proposed developments in the territories of the Mi'kmaq and the Indigenous peoples of the NWT are able to proceed, their world – and ours – will be forever harmed and lessened. This may be the most important lesson for us to learn: that we must address the "uniquely calamitous feature" of modern post-industrial culture, whereby we irrevocably harm the world we depend on for our future (Wildcat 2009, 39). One critical way to avoid this level of harm is to recognize the authorities of Indigenous peoples and accept the underlying validity of the reasons for so doing. An example from the United States sheds light on the value of acknowledging this authority. The US case of *City of Albuquerque v. Browner* addressed a dispute concerning the City of Albuquerque's waste treatment plant, which is located almost 10 km upstream of the Isleta Pueblo reservation lands. The court recognized the authority of this Indigenous group to adopt water-quality standards that are more stringent than federal standards and to implement those standards even when upstream point sources are located beyond tribal land.[4] Thus, the court recognized the right of the Isleta Pueblo people to regulate the City of Albuquerque's waste-treatment plant to protect their water source, even though it was upstream and off-reserve.[5]

While the basis and specifics of US law are certainly different from Canadian law, the underlying logic applies equally to downstream Indigenous peoples on both sides of the Canada-US border: meaningful Indigenous rights and responsibilities (especially those related to water) must include the right to challenge and limit detrimental upstream uses. However, Indigenous peoples choose their processes, which may include involvement in the existing structures of others, should these structures provide a meaningful opportunity to exercise Indigenous voices and create outcomes in line with Indigenous values, cultures, and rights. At the far other end of the continuum, when these structures fail, the law may provide some redress through legal challenges aimed at achieving social and environmental justice goals. The three case studies in this chapter fall somewhere in between these two ends of the continuum. They describe new processes and structures, each demonstrating a pathway toward achieving reconciliation in human relationships with water. Each involves Indigenous peoples and their governments choosing their own forms of governance regarding the water that flows through their territories.

In most Western conceptions, a border takes what is open and creates something closed. The challenge we face as a society in managing the way we use water is to overcome the conceptual limitations we have placed on ourselves, largely through our use and ideas about boundaries. Our strength does not reside in denying the entitlements and opportunities of Indigenous peoples. It is time to acknowledge that we will and always have found our greatest strength in cooperation. Cooperation that recognizes ecological limits and realities may be called ecological co-governance, a concept known to both Indigenous and non-Indigenous knowledge systems; we could apply it if we chose to look deeply into those teachings with a view to commonalities. To create a sustainable and ethical future, we need to cross the cultural boundaries we have created between us and work on the common space that resides there: the space of connection between ourselves, and between the lands and waters that sustain us and all our relations.

Acknowledgments

This article was written only because of the generous donations of time and insights given by Shelley Denny, Dan Christmas, Minister Michael Miltenberger, Emma Norman, Luanne Armstrong, and the Centre for Indigenous Environmental Resources (which gave me the time to write the article). Thank you to them all.

NOTES

1 This is a complex topic; see Karl S. Hele's (2008) text *Lines Drawn upon the Water: First Nations and the Great Lakes Borders and Borderlands* for a discussion on First Nation concerns in a uniquely international context.
2 For example, see the various water laws of each province that provide that all rights to and property in water are held by the Province and all uses require a license

granted by the Province. This excludes First Nations from involvement since First Nations are not decision-makers regarding water use.

3 R.V. Delgamuukw [1997] 3 SCR 1010.

4 Albuquerque v. Browner, 97 F.3d 415, 419, 422–24 (10th Cir.1996), cert. denied, 522 U.S. 965 (1997).

5 Ibid. at 419.

REFERENCES

Borrows, John. 2002. *Recovering Canada: The Resurgence of Indigenous Law*. Toronto: University of Toronto Press.

Burleson, Elizabeth. 2011. "Good Governance: Tribal, State and Federal Environmental Cooperation." In *Legal Strategies for Greening Local Government*, ed. Keith H. Hirokawa and Patricia Salkin, 207–53. Chicago: American Bar Association.

Chiefs of Ontario. 2006. "Submission to the Expert Panel on Safe Drinking Water for First Nations." Brantford: Chiefs of Ontario.

Collaborative Environmental Planning Initiative. 2010. "The Spirit of the Lakes Speaks – A Way Forward." http://brasdorcepi.ca/wp/wp-content/uploads/2011/07/Spirit-of-the-Lake-speaks-June-23.pdf.

Deloria, Vine, Jr. 2003. *God is Red*. Golden, CO: Fulcrum Publishing.

Denny, Shelley. 2009. *Eel Grass Habitat Maps of the Near-shore Environment of the Bras d'Or Lakes*. Eskasoni, Nova Scotia: Unama'ki Institute of Natural Resources.

Government of the Northwest Territories. 2010. *Northern Voices, Northern Waters: NWT Water Stewardship Strategy*. Yellowknife: Government of the Northwest Territories.

Hele, Karl S., ed. 2008. *Lines Drawn upon the Water: First Nations and the Great Lakes Borders and Borderlands*. Waterloo: Wilfrid Laurier University Press.

Lavalley, Giselle. 2006. "Elder Jake Swamp." In *Aboriginal Traditional Knowledge and Source Water Protection: First Nations' Views on Taking Care of Water*. Ottawa: Chiefs of Ontario and Environment Canada.

Marshall, Albert. 2011. "Healing and Two-Eyed Seeing." http://www.integrativescience.ca/Principles/TwoEyedSeeing/. Accessed 12 March.

Miller, Bruce. 1996–97. "The 'Really Real' Border and the Divided Salish Community." *BC Studies* 112 (Winter): 63–79.

Parfitt, Ben. 2010. *Fracture Lines: Will Canada's Water be Protected in the Rush to Develop Shale Gas?* Toronto: Program on Water Issues, Munk School of Global Affairs, University of Toronto.

Pinkerton, E. 1991. "Locally Based Water Quality Planning: Contributions to Fish Habitat Protection." *Canadian Journal of Fisheries and Aquatic Sciences* 48 (7): 1326–33. http://dx.doi.org/10.1139/f91-159.

Pinkerton, E. 1993. "Co-Management Efforts as Social Movements." *Alternatives* 19 (3): 33–8.

Pinkerton, E. 2008. "Integrating Holism and Segmentalism: Overcoming Barriers to Adaptive Co-Management between Management Agencies and Multi-Sector Bodies." In *Adaptive Co-Management: Collaboration, Learning, and Multi-Level Governance*, ed. Derek Armitage, Fikret Berkes, and Nancy Doubleday, 150–71. Vancouver: UBC Press.

Wildcat, Daniel R. 2009. *Red Alert: Saving the Planet with Indigenous Knowledge.* Golden, CO: Fulcrum Publishing.

Wood, Chris. 2010. "The Last Great Water Fight: The Battle for the Northern Headwaters of the Mackenzie River." *The Walrus* 7 (8): 26–36.

3 Rise of the Local? Delegation and Devolution in Transboundary Water Governance

EMMA S. NORMAN AND KAREN BAKKER

Protecting shared water resources requires a commitment to cooperation. In the conclusion of chapter 2, Merrell-Ann Phare stresses the importance of enabling cooperation by emphasizing issues that unite rather than divide users of shared waters. Cooperation is complicated, of course, by the existence of different political regimes, administrative constructs, and legal and jurisdictional frameworks. Accordingly, innovations in governance require significant changes to laws, governmental organizations, and decision-making processes, and also to our worldviews of our relationships to water, land, and political territories (see box 3.1). To some degree, such a shift has already taken place over the past several decades in North America: water governance has undergone a twofold rescaling: "down" from national governments to local decision-makers and "out" from government-led decision-making structures to nongovernmental actors. The increased involvement of local, nongovernmental actors is a feature of environmental governance more generally. But it is striking that it has occurred along the Canada-US border despite (as explored below) the geographical, jurisdictional, and political barriers that hinder cooperation between local actors and the fact that international waters have traditionally been the domain of federal governments.

Box 3.1 Yukon River Inter-tribal Watershed Council

Emma S. Norman

The Yukon River Inter-Tribal Watershed Council (YRITWC, or the Council) is a collective initiative of 70 First Nations and tribes across Alaska and the Yukon Territory that aims to improve the health and well-being of the watershed and the people who live within it. The Council's vision, simply put, is "to be able to drink water directly from the Yukon River" (YRITWC 2009).

The multijurisdictional (and transboundary) nature of the watershed, in years past, had complicated the governance of the watershed. While agencies at the federal, state, and / or territorial level had some regulatory responsibility for the watershed, no single group existed to manage the watershed in its entirety. Recognizing that need, the Council was established in 1997 as a treaty-based organization of indigenous governments dedicated to preserve and protect the environmental quality of the Yukon River for the health of their communities and the continuation of a traditional, native way of life for generations to come. The YRITWC is an innovative and highly collaborative organization, the first dedicated solely to promoting the responsible management, use, protection, and enhancement of the watershed. The Council achieves these goals through a variety of methods, including educational programs, water quality monitoring, stewardship, and land-management practices. In addition, the Council serves as a vehicle to involve the First Nations and tribal communities in direct decision-making related to the governance of the watershed and to provide a forum where member villages, tribes, and nations collectively can express their needs (YRITWC 2009). In 2005, Harvard University recognized the innovations of the Council as an award-winning program. The Council was described as a model of self-determination, governance, and collaboration, with high achievements in three main areas: the initiation of the YRITWC, the development of a complex and high-quality operational system, and the impact and reach of the Council on the health of Native peoples along the Yukon River and beyond (Harvard University 2005).

The Council continues to develop new programs with a focus of five main tenets:

1. Understanding the Watershed through monitoring, measuring, and researching, and using this knowledge to clean, enhance, and preserve life along the River.
2. Education: Promoting environmental and traditional education for the Indigenous Peoples of the Watershed through educational programs, scholarships, internships, volunteer opportunities, and incentive programs.
3. Stewardship: Honouring the traditional heritage through good stewards of the Watershed and its tributaries, and restoring and preserving its health for the benefit of future generations.

4. Enforcement: Developing and enforcing strong state, federal, territorial, and provincial environmental standards to preserve the long-term health of the Watershed.

5. Organization: Providing greater organizational strength to the Indigenous Peoples of the Yukon River Watershed (see Figure 3.1), both by assisting and improving Indigenous governments and by being a model of organization built on collaboration and mutual respect. (YRITWC 2009).

References

Harvard University. 2005. Innovations Award: Yukon River Inter-Tribal Watershed Council. http://www.innovations.harvard.edu/awards.html?id=16859.

YRITWC. 2009. Yukon River Inter-Tribal Watershed Council. http://www.yritwc.org/AboutUs/AboutUs/tabid/56/Default.aspx

Figure 3.1 Yukon River Watershed

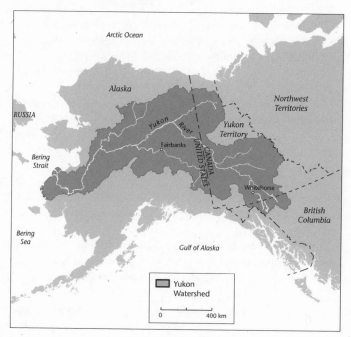

Source: Original Map by Eric Leinberger, University of British Columbia.

In this chapter, we provide a brief explanation of why local-level decision-making is increasingly important. We also examine the implications, because (in contrast to much of the literature) we do not begin from the assumption that local governance is necessarily better or more equitable. In particular, we discuss the capacity of the local (sub-state) actor to engage in decision-making processes, and we ask two questions: Are local actors empowered to make decisions? Do they have the capacity to carry out the initiatives they have been assigned (or taken on)? In our analysis, we find that the process of rescaling of water governance to the local level is not necessarily empowering for local actors. This conclusion is at odds with an assumption prevalent in much of the water governance literature, namely, that rescaling to the local level will be empowering and, moreover, that this empowerment will in turn lead to better water management outcomes. In our closing remarks, we offer suggestions for more meaningful engagement of local actors – who play a crucial role in governance of shared waters, but whose capacity for meaningful engagement has not yet been fully realized.

Managing Water at the Local Scale

Discussions in the water governance literature often highlight the benefits of addressing environmental issues at a local level (Lundqvist et al. 1985; Gleick 1993; IJC 2000; Kliot et al. 2001). A strand of environmental management literature says that the involvement of local actors tends to legitimize policy and environmental programs, described by Corry et al. (2004) as "new localism." The new localism advocates local involvement as necessary and positive, and as a means to replace higher-order levels of government and reinforce the emergence of "social trust" in which both public and private needs are met and local democratic institutions are enabled. From this perspective (one that we query, below), local governments are more in touch with community needs, more empowering, more effective in cooperative practices, and more cost-efficient than "higher" scales of governance.

The "new localism" is thus complementary to the dominant paradigm: Integrated Water Resources Management (IWRM), which asserts the importance – even primacy – of the local scale in the governance of water resources (Cohen and Davidson 2011; Biswas 2004; Blomquist and Schloger 2005). A common, and often implicit, presumption in the IWRM literature is that human, environmental, and social decisions should be integrated through water basin – based governance instruments, rather than through political jurisdictions. Until recently, few scholars have looked critically at how the concepts of watersheds are constructed, conceptualized, and applied in practice. But, as Cohen and

Davidson (2011) point out, there is great need to disentangle the concept of "local governance" from "watershed management" because watersheds are not necessarily "local" and often rely on established administrative structures that are not commensurate with the watershed scale.

We suggest that the implicit assumption about the positive benefits of down-scaling governance to the local level is problematic, for three reasons: (1) it implies an equitable inclusion of stakeholders across the basin; (2) watershed basins are not necessarily "local" (think, for example, of the Fraser or the Mackenzie watersheds) (see also Cohen 2012); (3) limited local capacity may hinder effective decision-making. Another potential pitfall is that "down-scaling" may, in a manner of speaking, downshift responsibility without the allocation of resources necessary to undertake newly assigned responsibilities. Brown and Purcell (2005) term this phenomenon "the local trap"; it is an analytical counterpart to Agnew's "territorial trap," which assumes that organization, policies, and action at the local scale are inherently more likely to have desired social and ecological effects than activities organized at other scales. As Cochrane (1986, 51) claims, "governments seem to use community as if it were an aerosol can, to be sprayed on any social programme, giving it a more progressive and sympathetic cachet."

A related point pertains to the *process* of rescaling governance. The literature rarely questions the redistribution of power that must occur alongside rescaling of governance process, which must occur if local actors become more central in water governance. Yet these rescaling processes may have significant implications for the ability of actors, at all levels and scales, to engage effectively in water governance. For example, as Part II of this volume demonstrates, rescaling may result in mismatches of administrative and governmental bodies between water basins, which would complicate the governance of transboundary waters.

In addition, calls for the watershed approach frequently imply that nation-states necessarily lose a degree of power and influence when transboundary waters are managed according to watersheds, and that devolution of responsibility and authority to local and supranational scales also implies an increasing porosity of borders to local actors (Coates 2004). In other words, embedded within the literature (and environmental management literature, more generally) are the implicit suppositions that transboundary watersheds emerge inevitably from processes of decentralization and that this rescaling necessarily increases the power of local scales at the expense of the nation-state. This raises the risk of treating the involvement of local actors in a relatively uncritical fashion, particularly with respect to assumptions of equitable and meaningful participation, significant influence over decision-making, and accountability or empowerment.

These critiques, to some degree, are anticipated, and certainly enriched, by conversations in borderlands studies, which show how the creation of a border is simultaneously material and symbolic. The idea that nations are fixed and territorially bounded spaces is questioned in borderlands literature (see also chapter 2, this volume). However, the materiality of the border impacts people differently, particularly as we investigate issues associated with power, space, and race (Norman 2013). Critically analyzing the process of border creation from a historical perspective, Paasi (1996) argues geopolitical borders reinforce national identity by physically keeping citizens in (and outsiders out). Nation-building narratives help citizens internalize and reify national identities. Other scholars have made similar arguments, calling for geopolitical borders to be situated within wider historical frameworks and to be recognized as socially constructed spaces of power (see for example, Fall 2005, Furlong 2006, and Popescu 2012). This trend is particularly important in transboundary discussions, which often neglect to question the inherent power relations in the boundary-making process. For example, nation-state boundaries have significant consequences for Indigenous communities. which predate the construction of many state borders (see chapter 2 this volume) and which continue to be impacted by contemporary border constructions and their associated laws and policies. This holds true particularly in relation to access to culturally relevant marine resources that transcend the borders (Thom 2010; Norman 2012).

Treating scales of governance as interconnected, constructed, and evolving, rather than distinct, naturalized, and immutable spaces, help us analyse these border regions in a more nuanced way. This approach is particularly relevant to resource sectors that have experienced significant rescaling of environmental governance in recent years. This, in turn, justifies an interrogation of the implications of rescaling for local communities and hydro-social cycles. It also implies scepticism concerning assumptions often embedded in prioritization of the watershed scale in water governance debates. Implementing watershed governance, for example, does not automatically imply equitable representation of all stakeholders or more power for local stakeholders vis-à-vis higher orders of government. Rather, this perspective reminds us of borders' simultaneous fixity and porosity, and documents how local actors simultaneously undermine, yet are constrained by, the container of the nation-state. Indeed, ironically, local actors are often less able to transcend the border than their nation-state counterparts.

In short, we query certain core assumptions prevalent in the transboundary water governance literature in order to better understand the needs to achieve more effective transboundary governance of water. In particular, we question

Table 3.1 Number of Federal and Subnational Governance Instruments Created per Decade (1900–2007)

	Federal		State-Provincial		Multilevel		Local		Total Subnational	
	Number	%	Number	%	Number	%	Number	%	Number	%
Binding	73	76.84	15	42.86	0	0.00	0	0.00	15	20.55
Nonbinding	3	3.16	12	34.29	2	6.45	1	0.17	15	20.55
Organization	19	20.00	8	22.86	29	39.55	5	0.83	42	57.33

Note: Total includes all 166 governance mechanisms, including organizations; federal and subnational exclude organizations.
Source: Canada-US Transboundary Water Governance Instruments Database (2007).

the notion that borders have become more porous over time, as well as the assumption that scaling downward to the local level implies that the border is more fluid; these topics are the focus of the following sections.

Rescaling Canada-US Transboundary Water Governance

In the preceding section, we explored presumptions related to the "local trap" in the environmental governance literature. To flesh out this critique, we now turn to the governance of transboundary water for the Canada-US border.

In this section, we analyse whether and how rescaling of transboundary water governance along the Canada-US border has occurred. The analysis draws on our comprehensive database of international, water-related governance mechanisms between Canada and the United States, managed at multiple scales: local, provincial and state, national, and international (see Norman and Bakker 2009 and www.watergovernance.ca for information on database). Table 3.1 explores the relationship between scales of governance and type of governance instrument. We define "instrument" as a device to govern water, such as a treaty, exchange of notes, memorandum of agreement, memorandum of understanding, agreement, order, and organization.

Our findings indicate a total of 166 water-related governance instruments, 57 per cent of which were federal and 43 per cent of which were subnational (state or provincial, multilevel, or local). When disaggregated into formal and nonformal (treaty and nontreaty), the federal instruments clearly rely more heavily on formal agreements (77 per cent), whereas subnational or multiple-scaled groups rely more on organizations (57 per cent) and informal agreements (16 per cent). This is unsurprising, given the limited capacity for local organizations to create "binding" or "formal" agreements in an international setting. In lieu of binding agreements, the subnational, particularly at the multiscale and local levels, rely heavily on organizations. Through the creation of groups dealing with a singular issue (such as the Flooding of the Nooksack River Task · Force in the Pacific region) or basinwide issues (such as the Gulf of Maine Council in the Atlantic region), these organizations are able to create networks with relatively little infrastructure to accomplish information sharing and problem solving. Overall, we found that the number of instruments designed to manage transboundary water has substantially increased over time, as has the rate of growth of new instruments since the early 1980s (Figure 3.2). The rate of growth of instruments was relatively slow in the first half of the twentieth century. From the 1940s through the 1970s, the growth in instruments stayed steady, averaging about 15 new instruments per decade, but through the 1980s, the number of new instruments doubled to 26 and then increased again to 37 during the 1990s. From 2001 to the present, 25 instruments have been established. This rate is congruent with the 1991–2001 rate, with an average of 3.6 instruments per year. Moreover, whereas the period up until the 1940s was dominated by treaties, agreements, and exchanges of notes, the majority of instruments created since the 1980s were conceived and implemented at the subnational level, as described later.

Figure 3.2 Number of federal and subnational governance instruments created per decade (1900–2007)

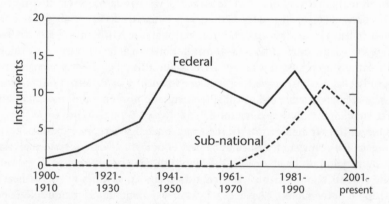

Source: Canada-US Transboundary Water Governance Instruments Database (2007).

Analysing the data temporally also reveals some very interesting trends regarding the changing roles of federal and subnational actors in water governance. Figure 3.2 shows a trend of declining federal involvement and increasing local involvement, in terms of the number of organizations and instruments. Even when evaluating solely governance instruments, excluding organizations, the trend clearly indicates a rise in local participation since the early 1990s. Our analysis reveals that the federal role peaked in the 1940s, during what Pentland and Hurley (2007) call the cooperative-development period (see Table 1.1, this volume). In fact, the trends found in our analysis coincide closely with the transboundary water periods identified by Pentland and Hurley (2007). As the federal role declines, the local-governance instruments start to emerge during the comprehensive-management era in the late 1960s and 1970s. Not until the 1990s, however, in the middle of the sustainable-development era and the beginning of what we call the participatory era, did the role of the local level reach its zenith. When analysing this trend before and after NAFTA (1992), a trend of greater local involvement emerges a decade after NAFTA was signed. This trend is significant given the role of NAFTA's environmental side agreement, which established the Commission for Environmental Cooperation (CEC), a trinational, transboundary environmental agency with funding mechanisms. Several of the recently established organizations, including multilevel groups in the Pacific and the Atlantic regions. received funds from the CEC.

These trends in water governance parallel trends reported by the IJC. In a recent Canadian Department of Foreign Affairs presentation, representatives from the IJC reported that the instruments used to enable the IJC, such as references and applications, have steadily declined over the past several decades (for a detailed discussion of the IJC, see box 1.1. and chapter 4, this volume). In an effort to adapt to these changing patterns, the IJC has developed a new mechanism in the form of the watershed approach, which is intended to allow for greater multilevel, and specifically local, participation (IJC 1997, 2000, 2005).

Tables 3.2 and 3.3 show how federal and subnational instruments are distributed spatially across border regions. The highest number of local instruments is found in the Pacific region; however, the Atlantic and Western Montane have the highest percentages of local instruments as a proportion of the total. The Great Lakes – St. Lawrence region has the lowest proportion of local instruments, which may result from the scale of the lakes, where fewer bodies of water means fewer instruments; a large number of actors, which complicates any local agreement; or the level of IJC involvement, which keeps activities focused at the federal scale. Despite the proportionally small amount of local-level participation, the Great Lakes region was one of the first to include multilevel stakeholders in transboundary water governance with the establishment of the

Great Lakes Fishery Commission in 1955. In terms of provincial – state relationships, the Pacific region led the way with the founding of the Environmental
Cooperation Council in 1992. This binational organizational body was emu-
lated in Montana – Alberta and British Columbia – Montana more than a decade later.

Table 3.2 Regional Variation of Governance Instruments

Basin Region	Country-Wide	Pacific	Western Montane	Central Prairie	Great Lakes	Atlantic
Federal						
Organization	2	3	1	7	5	1
Treaty	2	3	0	4	7	0
Agreement	1	3	0	0	7	3
MOU/MOA	1	0	0	0	2	0
Exchange of Notes	4	5	2	2	9	5
IJC Order	0	8	1	4	3	0
State-Provincial						
Organization	0	5	2	0	0	1
Agreement	0	5	2	0	5	3
MOU/MOA	0	4	2	2	1	3
Multi-Level						
Organization	0	8	2	7	6	5
MOU/MOA	0	0	0	0	1	1
Nongovernment						
Organization	0	2	1	0	0	2
MOU/MOA	0	0	0	0	1	0
Sub-Total	10	46	13	26	47	24

Source: Canada-US Transboundary Water Governance instruments database (2007).

Table 3.3 Percentage of Local Governance Instruments per Region

Region	Federal	Local	% Local/Total
Country-wide	10	0	0
Pacific region	22	24	52

(Continued)

Table 3.3 *(Continued)*

Region	Federal	Local	% Local/Total
Western Montane	4	9	69
Central Prairie	17	9	35
Great Lakes – St Lawrence	33	14	30
Atlantic	9	15	63

Source: Canada-US transboundary water governance instruments database (2007).

Another important aspect of our analysis examined the spatial clustering of governance instruments around watersheds, rather than larger regions. Through this analysis, we found that a few basins make up a large proportion of the governance instruments. We found that 61 per cent of the transboundary governance instruments were created for eight (out of 30) transboundary watersheds: Great Lakes, 23; St. Lawrence River, 17; Columbia River, 15; Niagara River, 10; Red River, nine; Rainy Lake, eight; Georgia Strait-Puget Sound, eight; St. Croix River, seven. In fact, the top five watersheds make up almost 50 per cent of the governance instruments. A significant amount of this activity occurred during the cooperative development period in preparation for shared hydroelectric development. In the Great Lakes region, though, which has the greatest number of governance instruments, the efforts have been both sustained and dynamic. The transboundary instruments date back from the beginning of binational cooperation with the Boundary Water Treaty and the International Lake Superior Control Board (1925), to the Great Lakes Water Quality Agreement (1972), to, most recently, the preparation for the Upper Great Lakes Study Board (2007). The second most active geographic region, the Pacific, looks notably different from the Great Lakes – St. Lawrence River region. In the Pacific, the development of water governance instruments has emerged largely during the 1990s in the sustainable development and participatory era. Leading the way in binational cooperation at the provincial – state level, the Pacific developed a significant amount of its instruments to work with many multistakeholders at various scales. A notable exception is the Columbia River, where many of the binational governance instruments were developed in 1940s to support its role in generating hydroelectric power and managing flood control. Like the Great Lakes, however, the governance instruments are dynamic, with efforts to become more participatory in recent times (e.g., the Columbia Basin Trust [est. 1995] and the Columbia River Transboundary Gas Group [est. 1998]).

Porous Borders?

In the preceding section, we documented the reconfiguration of governance instruments for transboundary water. The trend of a declining federal role and an increasing local role was clearly exhibited (Figure 3.2). We also found significant regional variances in governance instruments across the border, where a relatively small number of watersheds comprised the majority of governance instruments. We now query whether this reconfiguration has led to increased porous borders and empowerment (increased institutional capacity and involvement in decision-making processes) for local actors. Although the majority of respondents identified an increased participation of local actors in water governance over the past 15 years, they consistently articulated the opinion that this increase failed to translate into greater local institutional capacity. Nor has it enhanced their ability to travel across the Canada-US border (particularly after 11 September 2001).

Several of the respondents, however, noted that this trend should not be simplistically framed as uniformly positive; indeed, in many cases, rescaling is the result of a downloading of responsibility by senior government. For example, the Columbia River in Washington and British Columbia was identified as an example where a shift from federal responsibilities to mixed arrangements between state, province, NGOs, and tribal governments has occurred with mixed results. A participant in one focus group reflected:

> Yes, local involvement and importance has increased, *but* the greater good has to be considered. We cannot let "single-issue" groups make decisions. However, local stakeholders must be involved or else things just don't happen.

Similarly, a second focus group noted that "local governments are more susceptible to local political pressures, such as land development, if there is no state or provincial standard to be met." These findings are not unique to the Canada-US border. In a recent study including 83 river-basin organizations worldwide, it was found that, although decentralization to the "lowest appropriate level" is an internationally accepted principle of river-basin management, the "actual application often encounters obstacles due to the varying interests of different stakeholder groups" (World Bank 2007; see also Cassar 2003). Our empirical work explores the root causes for the lack of increased capacity for transboundary water governance at the local scale. Our analysis indicates that mismatched or asymmetrical governance structures, limited institutional capacity, and lack of intrajurisdictional integration all play an important role in limiting the extent and effectiveness of transboundary cooperation. The continued emergence of flashpoints along the borders underscores the implications of these barriers.

Table 3.4 provides a list of drivers and barriers of transboundary cooperation identified through interviews and focus groups. Respondents also argued that asymmetrical governance structures at the local scale were aggravated by the different pace and timing of rescaling in Canada and the United States. Specifically, in the United States, interview respondents indicated that despite statewide programming aimed to include local stakeholders in water governance activities, such as the 1987 *Water Act* Amendments, "water governance has decidedly *not* shifted to local communities." Some respondents stated that "there was actually *less* power for the local communities today than fifteen years ago." Although the 1987 amendments aimed to bring more power to local communities through state empowerment, some water managers felt that the "local communities had their hands tied by the amendments" and that "the idea that local communities have more power is [simply] illusionary."

Several of the US respondents further indicated that state employees tend to have little involvement in the transboundary process; many noted that transboundary governance "was either local or federal." Furthermore, many of the state employees felt that their hands were tied in terms of involvement in

Table 3.4 Transboundary Cooperation: Drivers and Barriers

Drivers	Barriers
Specific issues	Mismatched governance structures
Leadership	Different governance cultures
Informal contacts	Different mandates
Established networks	Lack of institutional capacity
Crisis	Lack of financial resources
Personal relationships	Asymmetrical Participation
Public availability of data	Data, lack of / difficulty accessing
Proximity	Lack of intrajurisdictional integration
Legal obligations	Gaps in knowledge of the 'other' country
Opportunity	Spatial distance
Transparency	Federal jurisdiction tempers regional action
Practicality	Mistrust
Respect / fairness	Lack of leadership

Source: Norman and Bakker (2009).

transboundary water issues. As noted by one respondent: "They [the feds] limit our opportunity – we *could* get involved, but then they [the feds] could just take over." For the local groups and environmental NGOs in the study region, we found that limited financial resources and overextended staff tempered binational cooperation. One of the difficulties was the constituent base, which was often reticent to have its donor dollars stretch across international jurisdictions. These barriers tend to limit the scale and scope of the groups' projects and limit their involvement in transboundary governance in general. In Canada, in contrast, jurisdictional fragmentation has led to confusion over appropriate roles and appropriate scales of responsibility for water governance. Our interviews echo the observations from Parson (2001, 131) that this fragmentation is exacerbated by a process of rescaling in which "environmental authority is being ceded at once downward to the provinces and upward to international institutions" (see also Paehlke 2001). In particular, the private sector and local governments have experienced an increase in responsibilities because of this devolution. Several of the interviewees described how this devolution of provincial authority has led to greater local participation in water governance, as well as greater federal responsibilities. In the words of one provincial respondent, "The local is becoming more responsible for water governance issues as a flow-over from provincial downsizing."

This reconfiguration has impacted volunteer organizations and Canadian NGOs, whose increased participation in environmental governance is well documented (Gibson 1999; Dorcey and McDaniels 2001; Howlett 2001; Savan et al. 2004), and growing influence in environmental policymaking and monitoring is recognized (Savan et al. 2003). Several provincial employees concurred that "more and more people are becoming involved in water governance issues" and that "there has been an increase in cooperation at the local neighbor – neighbor level." This participation has its limits, though, as one interviewee reflected: "Canada has a strong government-to-government mentality, which provides a barrier for citizen participation in transborder issues." Thus, although NGOs are more present and have increasingly more influence, the devolution of power has not yet translated to direct governing authority for nongovernmental actors. These observations are important for several reasons. First, they decouple the "hollowing out of the state" from "glocalization" by showing that an increase in local and nonstate participation does not necessarily lead to a reduction in federal power (despite declining federal involvement). Second, they speak to the "local trap" by showing that an increase in local involvement does not equate to increased power of the local role. Specific examples from our case studies, both in our interviews and focus groups, help flesh out these trends.

Our case studies reveal that the Canada-US border remains a significant barrier to governing shared water resources, despite the programming and energy expended toward transboundary governance at a local level. Several respondents attributed this lack of fluidity, or ability to move easily across the border, to limited institutional capacity. Issues such as inability to make phone calls internationally, travel across borders, purchase data, and generally work with counterparts contributed to this phenomenon. For example, one respondent noted that despite working on an international river basin for several years, until very recently, he or she was unable to call out of the country – the office was able to receive international calls but could not make them. This type of limited capacity also occurred with data acquisition. One interviewee highlighted the nuisance in getting departmental funds to purchase data from Canada pertinent to work on an international river basin. Because "the data was not public domain" and the department "was not quick to spend the money" to acquire the information, the respondent reported that he or she was forced to "bypass the organization and purchase the data using personal funds." Another respondent reflected on how intractable binational governance could seem:

> When I ask if I can get data, they [the Canadians] say "you can buy it or we don't have it." Then, I find out later they really *do* have it, but just didn't share it or they didn't know that it existed. It is frustrating .... In Alberta, they have great Web sites, very flashy and looks nice, but the raw data is not there – it doesn't seem to be easily available .... We do things differently on *our* side. We have more public information – USGS has a public website where you can get any information – it is free public information for anyone at anytime to access.

Similar issues with data were reported by the coastal Pacific respondents. As one interviewee noted:

> There is a lot more capacity in the United States [for water governance]. Even from the data perspective, they [the US] have more data to work with.

General lack of knowledge of one another was also consistently reported as a barrier to transboundary cooperation. Many of the water managers interviewed were largely unfamiliar with the political structure and key environmental legislation of the other country. They were, however, interested in knowing this information. Not only were the interviewees unfamiliar with the political structures of the counterpart's country, they were also largely unconnected to their counterpart across the border, despite the shared watersheds and water-related concerns. One binational creek in the Pacific region, for example, which is experiencing a significant decline in salmon population due to habitat loss and lack of water flow, has little coordination among agencies and groups managing the watershed. Despite the physical proximity of the border towns (approximately five kilometres), the staff (city, state, and NGO) that work on the management of the creek operate with little to no communication with their counterparts across the border. One extension officer from a Washington state agricultural office reported a desire to work with her or his counterpart but was unsure who that person was or which office to contact. In fact, the extension agent turned to the authors' project as a way to connect with his or her counterpart. In this case, the border is both materially and socially constructed – although the physical space connecting the creek is minor, the border creates a very real socially binding arena that provides barriers to coordination. Coordination between counterparts is even more strained when great geographic distance separates the water issues. Several of the interviewees identified distance as a significant factor limiting binational cooperation.

The St. Mary and Milk Rivers in Alberta and Montana were identified as an example where great distance between the issue and a geographically dispersed population limited civic engagement (see chapter 8, this volume). In this case, the sparse population at the border, particularly in Montana, has kept the issue largely in the hands of state and federal officials. More generally, the role of distance decay can partially explain the clustering effects around specific watersheds, such as the Great Lakes, Columbia, and St. Lawrence basins that receive the bulk of attention, whereas other transboundary water issues including the Skagit and Yukon Rivers receive much less. The bodies of water that are more visible at the national level tend to have more governance instruments built around them.

Inconvenient meeting venues were also identified as a limiting factor for civic engagement and a barrier to fluid borders. This limitation is exemplified by one local stakeholders' group in the Pacific region, where the meeting venue perpetuated significant asymmetry in participation. Although the group's mandate is binational (British Columbia and Washington), the meetings are almost always held in British Columbia, leading to significantly more participants from Canada than from the United States. The official roster has an equal number of participants listed; but only one member from the United States attended over the past six meetings (with an average participation rate of 12 people). To help mitigate this asymmetry, the group leaders agreed to meet at the border. Although the change in venue provided a neutral setting, minimized the amount of travel for the US participants, and eliminated the need to cross through customs, the asymmetry continued: only one US participant attended.

The difficulty in convening a binational forum serves as an example of the simultaneous fixity and porosity of borders. In this case, the local actors undermine the border by treating a binational watershed that spans a political border as a singular "borderless" bioregion. Due to physical constraints of the contained nation-state, however, the bordered region impacts the local actors, who are unable to transcend the border for their meetings. Mismatched political structures and governance mechanisms in Canada and the United States were also repeatedly identified as a barrier to transboundary cooperation. In a focus group survey of barriers to transboundary water governance, asymmetry consistently rose to the forefront of the issues. One respondent lamented,

> There is no symmetry to how decisions are made across the border. The fact that the US and Canadian governments have inverse state-federal power distributions significantly impacts how decisions are made. Because Canada has a strong provincial system and a weak federal system, and because the United States has a strong federal system and a weak (relative to provinces) state system, negotiations between counterparts are often tenuous.

Furthermore, different funding cycles and legal structures accentuate the difficulty in coordinating projects across the border. One respondent reflected,

> Because there are potentially four relevant jurisdictions (two federal, state, and provincial) and four different fiscal schedules to coordinate, it can be a nightmare to figure out .... [we try to] come up with ideas, implement pressure, and spend it all by the end [of the fiscal year] to ensure that there will be funding again next year.

This asymmetry of governance mechanisms further complicates the possibilities of transborder, basinwide management of water, particularly at the local level. It has been argued that the subnational organizations are more flexible than the more formal federal institutions, but in a transboundary setting, we find that the asymmetries in governance structures serve to immobilize, or at least temper, the ability for regional groups to work effectively across national boundaries. Conversely, the IJC was specifically developed to mitigate these asymmetries by creating a level playing field between nations. As the trend toward more localized scale persists, however, the IJC is receiving fewer references and applications for study.

The majority of our respondents argued that that cross-border cooperation was considered intermittent, issue driven, and unequal across the border.

The interviews reinforced our earlier findings that transboundary governance instruments are narrowly concentrated among a few major basins, rather than spread equitably across the border. As one provincial employee noted, the "higher profile watersheds and larger bodies of water tend to be the focus of transboundary committees (i.e., Columbia River, Georgia Strait – Puget Sound)." Several lower-profile water systems, although rife with environmental concerns, are eclipsed by these more public issues. We also found that the attention to transboundary watersheds was largely sporadic. This is true particularly for the lower-profile watersheds, for which issues arise mostly in times of crisis. As one senior Washington state employee reflected,

> It would be nice to have ongoing [transboundary] institutional cooperation. It [cooperation] has always been episodic .... We don't even have good interstate agencies. How are we supposed to have good international agencies if we can't even coordinate between states?

These points highlight our finding that despite the increase in the number of local – and decline in new federal – instruments to govern transboundary water, the nation-state remains a key instrument in negotiating transnational water issues. In other words, rescaling is not leading to a hollowing out of the state, nor to a new localism. Rather, the rescaling of governance instruments might better be described as a game of musical chairs, where the players might be changing, but the balance of power has remained relatively constant.

Conclusions: Querying the Power of Locals in Transboundary Water Governance

Our analysis speaks to recent debates over the rescaling of water governance. Above, we documented the rise of local participation as a key trend in Canada-US transboundary water governance, and explored some of the potential contributions and pitfalls of "rescaling" to the local level. This allowed us to document the constraints to which local transboundary water governance is subject and to question the desirability of strategies – currently being explored by policymakers – of giving greater weight to local transboundary water governance at a watershed scale between Canada and the United States. Specifically, our analysis has suggested that although rescaling of transboundary water governance has occurred (for instance, local actors are increasingly present in the rosters of transboundary governance decision-making bodies and processes), greater empowerment (specifically defined as institutional capacity) for local actors has not necessarily resulted. Indeed, ironically, we found that local actors are less able to act effectively across the border than their nation-state counterparts. Although local actors are genuinely attempting to engage in transboundary governance, they encounter limited success due to inadequate resources and restricted capacity. Thus, despite the documented increase in participation of subnational actors in transboundary water governance, significant barriers have limited the capacity for these actors to participate effectively in decision-making on the management of water resources across an international border. These barriers include asymmetrical participation, mismatched governance cultures and structures, spatial distance, and limited capacity. These findings are at odds with much of the water management literature, but they correspond with Fischhendler and Feitelson's (2005) argument that reducing the scope of transboundary governance to binational (i.e., federal government) actors is a successful strategy because it minimizes external players and lowers the political costs.

Our analysis thus suggests that success is by no means straightforward, bringing into question the current policy preference for downscaling. In addition, our analysis refutes the assumption – prevalent in the environmental management literature – that the rescaling of transboundary issues implies that the border is more porous and less fixed (Gibbins 2001; Corry et al. 2004; O'Riordan 2004). Rather, this analysis documented the relative fixity of the Canada-US border and explored how the asymmetrical governance structure and disparate governance rescaling trends in Canada and the United States limited managers' abilities to govern water across political borders. The nation-state retains key powers and authority to govern. Indeed, the federal scale offers the best hope of a "level playing field" between asymmetrical actors. This, in turn, suggests that we must reexamine the desirability and feasibility of local transboundary governance as the primary means of governing shared waters.

In closing, our analysis suggests two more general hypotheses for consideration in studies of the rescaling of environmental governance. First, the process of rescaling is not necessarily positive or empowering for its supposed beneficiaries. Rescaling may become a "downloading" of responsibilities without commensurate power and resources. In some cases, this is accidental; in others, intentional. Second, all scales are socially constructed – even apparently "natural" scales such as the watershed. The extent to which these scales are meaningful bases for social action rests on the shifting power geometries of actors at multiple overlapping scales. Analysing the activities of solely local actors or privileging any one scale in environmental governance risks misinterpreting the degree to which the processes of rescaling actually empower local actors.

In the chapter that follows, Clamen explores the role of the International Joint Commission's new International Watersheds Initiative. The creation of this new administrative unit is based on many of the assumptions discussed above. Our chapter has raised issues that are important to consider when assessing this new IJC initiative; for example, the conflation of "local participation" with "capacity" may lead to the (false) assumption that greater local participation will automatically enable inclusive, meaningful local water governance processes. Given our support (in principle) for more inclusive governance, and for the involvement of local actors – mitigated by our empirical findings regarding the constraints on local actors – it is our hope that our critique will open up positive dialogue on this topic.

Acknowledgments

This is based, in part, on the article that first appeared in the Annals of the Association of American Geographers 9 (1). Thank you to Routledge Press for the permission to publish part of this work in this edited volume.

References

Biswas, Asit K. 2004. "Integrated Water Resource Management: A Reassessment." *Water International* 29 (2): 248–56. http://dx.doi.org/10.1080/02508060408691775.

Blomquist, William, and Edella Schlager. 2005. "Political Pitfalls of Integrated Watershed Management." *Society & Natural Resources* 18 (2): 101–17. http://dx.doi.org/10.1080/08941920590894435.

Brown, C.J., and M. Purcell. 2005. "There's Nothing Inherent about Scale: Political Ecology, the Local Trap, and the Politics of Development in the Brazilian Amazon." *Geoforum* 36 (5): 607–24. http://dx.doi.org/10.1016/j.geoforum.2004.09.001.

Cassar, A. 2003. "Transboundary Environmental Governance: The Ebb and Flow of River Basin Organizations." In *World Resources 2002–2004: Decisions for the Earth: Balance, Voice, and Power,* ed. World Resources Institute, 158–9. Washington, D.C.: World Resources Institute.

Coates, B.E. 2004. "Is a Heftier Administrative Law Required to Prevent State Hollowing? The Case of Transborder Water Pollution." *International Journal of Sociology and Social Policy* 24 (1/2): 103–23. http://dx.doi.org/10.1108/01443330410790984.

Cochrane, A. 1986. "Community Politics and Democracy." In *New Forms of Democracy,* ed. D. Held and C. Pollitt, 51–77. London: Sage.

Cohen, A. 2012. "Rescaling Environmental Governance: Watersheds as Boundary Objects at the Intersection of Science, Neoliberalism, and Participation." *Environment & Planning A*: 44 (9): 2207–2224.

Cohen, A., and S. Davidson. 2011. "The Watershed Approach: Challenges, Antecedents, and the Transition from Technical Tool to Governance Unit." *Water Alternatives* 4 (1): 1–14.

Corry, D., W. Hatter, I. Parker, A. Randle, and G. Stoker. 2004. *Joining Up Local Democracy: Governance Systems for New Localism New Local Government Network, London.* London: New Local Government Network.

Dorcey, A., and T. McDaniels. 2001. "Great Expectations, Mixed Results: Trends in Citizen Involvement in Canadian Environmental Governance." In *Governing the Environment: Persistent Challenges, Uncertain Innovations*, ed. E.A. Parson, 247–302. Toronto: University of Toronto Press.

Fall, Juliet. 2005. *Drawing the Line: Nature, Hybridity and Politics in Transboundary Spaces.* Burlington, VT: Ashgate.

Fischhendler, Itay, and Eran Feitelson. 2005. "The Formation and Viability of a Non-Basin: The US-Canada Case." *Geoforum* 36 (6): 792–804. http://dx.doi.org/10.1016/j.geoforum.2005.01.008.

Furlong, Kathryn. 2006. "Hidden Theories, Troubled Waters: International Relations, the 'Territorial Trap', and the Southern African Development Community's Transboundary Waters." *Political Geography* 25 (4): 438–58. http://dx.doi.org/10.1016/j.polgeo.2005.12.008.

Gibbins, Roger. 2001. "Local Governance and Federal Political Systems." *International Social Science Journal* 53 (167): 163–70. http://dx.doi.org/10.1111/1468-2451.00305.

Gibson, R. B., ed. 1999. *Voluntary Initiatives: The New Politics of Corporate Greening.* Peterborough, ON: Broadview.

Gleick, P. 1993. "Water and Conflict, Fresh Water Resources and International Security." *International Security* 18 (1): 79–112. http://dx.doi.org/10.2307/2539033.

Howlett, Michael. 2001. "Complex Network Management and the Governance of the Environment: Prospects for Policy Change and Policy Stability Over the Long Term." In *Governing the Environment: Persistent Challenges, Uncertain Innovations*, ed. E.A. Parson, 303–46. Toronto: University of Toronto Press.

IJC. 1997. "The IJC and the 21st Century." Washington, D.C.: International Joint Commission. http://www.ijc.org/php/publications/html/21ste.htm.

IJC. 2000. "Transboundary Watersheds." Washington, D.C.: International Joint Commission. http://www.ijc.org/php/publications/pdf/ID1563.pdf.

IJC. 2005. "A Discussion Paper on International Joint Commission and Transboundary Issues." Washington, D.C.: International Joint Commission.

Jessop, Bob. 2004. "Hollowing Out the 'Nation-State' and Multilevel Governance." In *A Handbook of Comparative Social Policy*, ed. P. Kennett, 11–25. Cheltenham, UK: Edward Elgar Publishing.

Kliot, N., D. Shmueli, and U. Shamir. 2001. "Institutions for Management of Transboundary Water Resources: Their Nature, Characteristics and Shortcomings." *Water Policy* 3 (3): 229–55. http://dx.doi.org/10.1016/S1366-7017(01)00008-3.

Lundqvist, J., U. Lohm, and M. Falkenmark. 1985. "River Basin Strategies for Coordinated Land and Water Conservation." In *Strategies to River Basin Management*, ed. J. Lundqvist, U. Lohm, and M. Falkenmark, 5–17. Dordrecht: D. Reidel Publishing Company. http://dx.doi.org/10.1007/978-94-009-5458-8_2.

Newman, David, and Anssi Paasi. 1998. "Fences and Neighbours in the Postmodern World: Boundary Narratives in Political Geography." *Progress in Human Geography* 22 (2): 186–207. http://dx.doi.org/10.1191/030913298666039113.

Norman, Emma. 2012. "Cultural Politics and Transboundary Resource Governance in the Salish Sea." *Water Alternatives* 5 (1): 138–60.

Norman, Emma S. 2013. "Who's Counting? Spatial Politics, Ecocolonisatin, and the Politics of Calculation in Boundary Bay." *AREA* (Royal Geographic Society). http://dx.doi.org/10.1111/area.12000.

Norman, Emma S., and Karen Bakker. 2009. "Transgressing Scales: Water Governance across the Canada – U.S. Borderland." *Annals of the Association of American Geographers. Association of American Geographers* 99 (1): 99–117.

O'Riordan, T. 2004. "Environmental Science, Sustainability and Politics." *Transactions of the Institute of British Geographers* 29 (2): 234–47. http://dx.doi.org/10.1111/j.0020-2754.2004.00127.x.

Paasi, Anssi. 1996. *Territories, Boundaries and Consciousness: The Changing Geographies of the Finnish-Russian Border*. Hoboken, New Jersey: Wiley.

Paehlke, Robert. 2001. "Spatial Proportionality: Right-Sizing Environmental Decision-Making." In *Governing the Environment: Persistent Challenges, Uncertain Innovations*, ed. E.A. Parson, 73–125. Toronto: University of Toronto Press.

Parson, Edward A. 2001. "Environmental Trends: A Challenge to Canadian Governance." In *The Environment: Persistent Challenges, Uncertain Innovations*, ed. E. A. Parson. Toronto: University of Toronto Press.

Pentland, Ralph, and Adele Hurley. 2007. "Thirsty Neighbors: A Century of Canada-U.S. Transboundary Water Governance." In *Eau Canada: The Future of Canada's Water*, ed. K. Bakker, 163–84. Vancouver: UBC Press.

Popescu, Gabriel. 2012. *Bordering and Ordering the Twenty-first Century*. New York: Rowman and Littlefield.

Savan, Beth, Christopher Gore, and Alexis J. Morgan. 2004. "Shifts in Environmental Governance in Canada: How Are Citizen Environment Groups to Respond." *Environment and Planning. C, Government & Policy* 22 (4): 605–19. http://dx.doi.org/10.1068/c12r.

Savan, Beth, A.J. Morgan, and C. Gore. May 2003. "Volunteer Environmental Monitoring and the Role of the Universities: The Case of Citizens' Environment Watch." *Environmental Management* 31 (5): 561–8. http://dx.doi.org/10.1007/s00267-002-2897-y. Medline:12719888

Thom, B. 2010. "The Anathema of Aggregation: Towards 21st-century Self-government in the Coast Salish World." *Anthropologica* 52: 33–48.

World Bank. 2007. "Integrated River Basin Management and the Principle of Managing Water Resources at the Lowest Appropriate Level – When and Why It Does (Not) Work in Practice?" In *Global River Basin Management Research Project*, ed. World Bank. Washington, D.C.: World Bank.

4 The IJC and Transboundary Water Disputes: Past, Present, and Future

MURRAY CLAMEN

Created by the Boundary Waters Treaty of 1909 (BWT, or the Treaty), the International Joint Commission (IJC, or the Commission) is regarded by most water professionals, academics, and governments at all levels as having had a long and distinguished record of achievement. The establishment of the IJC more than 100 years ago provided an early institutional arrangement for international cooperation in the management of boundary and transboundary waters. Together, the BWT and the IJC provided the assurance of a shared system of principles and values (see boxes 1.1, 1.2, and 1.3), and clear mechanisms for dealing with uncertainty. This chapter deals with what is arguably the most important element of the BWT: the creation of the IJC.

Throughout its 100-year history, the Commission has dealt with about 120 cases. Although most of these relate to nation-to-nation disputes over water, a small number involved air and land issues, and, most recently, have included more public and inclusive forms of decision-making. In its first 50 years, the Commission dealt with many applications for works affecting boundary water levels and flows, and commissioners primarily took an engineering and legal perspective. Many of the projects the IJC reviewed and approved were related, for the most part, to hydroelectric power and irrigation. Conversely, in the last 50 years of its existence, the IJC has extended its work to include a much broader scope of issues: water levels and flows (in the Great Lakes and elsewhere), water and air pollution, river-basin development (in the Columbia and elsewhere), and water apportionments have all fallen under the umbrella of the IJC's work.

This variation reflects a number of factors, including general changes in societal values, the way in which the public interacts and expects to be involved in public-policy issues, and the Commission's ability to learn and adapt from its own experiences. These inclusions are examples of the IJC's institutional flexibility – flexibility that, I argue, has been central to the ongoing relevance and utility of the Commission more generally.

In this chapter, I show how the IJC's institutional flexibility has been central to its ongoing success. To do this, I draw primarily on one example: the development of the IJC's International Watersheds Initiative (IWI). The IWI was first proposed as a collection of international watershed boards in 1997 and is still in its early stages. The development of the program – from the initial idea through to the pilot projects – is a prime example of the kind of flexibility that has allowed the IJC to flourish through more than 100 years of changes in political, economic, and social climates. Joint fact-finding, objectivity, independence, and a forum for public participation are other attributes of the IJC that have contributed to its success. In addition, the creation and development of the IWI illustrates the fact that the IJC (and transboundary water governance in general) is at a crossroads in terms of meeting the environmental challenges of the twenty-first century within the framework of the century-old Boundary Waters Treaty.

After a century of addressing many issues arising under the BWT, the evolution to international watershed boards by the IJC (and the broader International Watersheds Initiative that is driving its implementation) is, in my view one, of the most exciting new concepts in transboundary environmental governance and holds great promise to help prevent and resolve transboundary disputes in the next century. Successful implementation is requiring the IJC to reconsider the BWT's essential purpose, as well as new and emerging natural-resource management trends in and between the United States and Canada. While several aspects of the IJC process could have been selected to illustrate the adaptability, flexibility, and robustness of this historic institution, I believe the watershed boards concept and the IWI best exemplify these traits.

This introspection is intended to illustrate the key challenges the IJC faces and the actions it has taken that might be transferable to other governance issues, and perhaps even to other parts of the globe. As a career IJC employee with more than 30 years of experience serving as an engineering adviser and secretary in the Canadian section of the IJC, I am in a unique position to report on these efforts from having worked with so many commissioners and personally experienced most of the events described. Before proceeding, I wish to reiterate my highest esteem for the IJC as an institution, as well as for the individual commissioners who make up its core. Throughout the commissioners' deliberations in private executive sessions, I have almost always observed that commissioners strongly desire consensus and want to respect the expert advice provided by their scientific boards, the guidance of the Treaty, and the concerns of the interested public. That is not to say, however, that the IJC does not have its critics. Some argue the IJC has become too political and is now not living up to its potential as an independent watchdog of governments, especially under the Great Lakes Water Quality Agreement. Others complain that the IJC takes too long to issue its reports, and some maintain those reports always take an "environmental" perspective and recommend against projects the Commission has been asked to study. Some suggest that the Treaty is outdated and should be amended; others, that new organizations can do the work of the IJC faster, better, and cheaper. This chapter will not deal with these criticisms, but instead will try to illustrate the organization's flexibility and adaptability in the way it adopted an ecosystem and watershed perspective through one new and important initiative: international watershed boards and the International Watersheds Initiative.[1]

This chapter is outlined as follows. First, I trace the origins of the concept of international watershed boards at the IJC and summarize the Commission's subsequent activities. I then outline the challenges the IWI has faced and identify the initiatives' successes. I reinforce these with case studies of our pilot projects and close the chapter by highlighting key elements of the IWI success. I urge readers to see this chapter as an examination not only of the IWI, but also of the IJC as a whole and the processes, challenges, and opportunities afforded by a 100-year-old treaty.

Origin of the International Watershed Board Concept and the International Watersheds Initiative

The idea for international watershed boards was first introduced in the IJC's 1997 report, *The IJC and the 21st Century* (IJC 1997). This was a document produced in response to an April 1997 request of the governments of Canada and the United States (Minister of Foreign Affairs 1997) to advise them on "how the Commission itself might best assist the parties to meet the environmental challenges of the twenty-first century within the framework of their treaty responsibilities." This request, or "charge" as the commissioners called it, arose first as an idea stemming from informal discussions between IJC commissioners and government officials on how to mark the 90th anniversary of the 1909 Treaty and the IJC, and second as an opportunity for an "announceable" for a meeting between Prime Minister Jean Chretien and President Bill Clinton.

Strictly speaking, the request was a reference under the Treaty, but was quite different from almost all previous references, since it focused on the Commission itself and not a dispute of any kind. Commissioners were quick to recognize the opportunity this request presented, and they worked to build on the extensive experience of the IJC to suggest some novel ideas for addressing transboundary water issues in a nonconfrontational way. After extensive consultations in both countries, the Commission put forward five recommendations. The recommendation on which the most progress has been made is that concerning the formation of international watershed boards. The idea for such boards was, in many ways, a reflection by the Commission of the ecosystem approach advocated and implemented in the Great Lakes under the various Great Lakes Water Quality Agreements (IJC United States and Canada 1972, 1978, 1987).

The governments responded to the Commission's international watershed boards proposal, in a letter dated 19 November 1998, with what was a new and different reference (Minister of Foreign Affairs 1997). They supported the concept and asked the Commission, in consultation with relevant stakeholders, to "further define the general framework under which watershed boards would operate"; make detailed recommendations on the location, structure, and operation of the first watershed board; and identify and plan for additional watershed boards. This 1998 reference has become an ongoing one pursuant to which the Commission has issued three reports to date (IJC 2000, 2005, 2009).

The International Watersheds Initiative was a reformulation of the international watershed boards concept and was introduced as a way to promote a new approach to transboundary water management. Rather than dealing only with high-level government action on water, the watershed approach introduced an integrated ecosystem approach (i.e., interacting components of air, land, water, and living organisms, including humans) through enhanced local participation and strengthened local capacity. The underlying premise was that local people, given appropriate assistance, were those best positioned to resolve many local transboundary problems. A key conclusion of the Commission in its 1997 report was:

In the past, transboundary water issues were often seen as localized at a specific dam or structure, or were examined as pollution problems in isolation from other factors. Experience with the Great Lakes Water Quality Agreement and the ecosystem approach [has] changed that perspective. Transboundary waters must be addressed in an integrative manner, including both biophysical and human aspects. … the new international watershed boards would adopt an integrative, ecosystem approach to the full range of water-related issues that arise in the transboundary environment, including consumptive uses, diversions and effects of air deposition and volatilization. (IJC 1997)

What Is the Watershed Approach?

The watershed approach is a strategy based on three pillars: an ecosystem approach to environmental management, the importance of integration, and the value of public participation in environmental decision-making (Cohen and Davidson 2011). Like many (CWRA 2004; International Joint Commission 2009; WMO 2010), the IJC believes this approach will lead to better and more effective water governance. The three pillars of this approach are briefly outlined here.

Ecosystem approach

According to one definition (WRI 2000) quoted in the 2009 IJC report, an ecosystem approach "broadly evaluates how people's use of an ecosystem affects its functioning and productivity," and:

- Considers the entire range of goods and services that can be derived from the environment
- Recognizes that ecosystems function as whole entities and should be managed as such
- Takes a long-term view
- Maintains the productive potential of ecosystems, seeking to preserve or increase their capacity to produce desired benefits in the future

In the case of water, an ecosystem approach means working within a topographically defined hydrologic drainage area, or an area of land draining into a common body of water. This definition includes not only the main water body, but also its tributaries and the land around it, which is a major change for the IJC.

Integration

Flowing from the ecosystem approach is the idea of integration. Given that no one component of the socio-environmental landscape functions in isolation, the watershed approach supports management initiatives that recognize, reinforce, and address the interconnections between the myriad elements of watersheds. These interconnections include, but are certainly not limited to, the integration of water quality and quantity, cooperation and coordination between multiple governments and their departments and ministries, and the interrelationships between land and water management.

Public Participation

The IJC believes that transboundary watershed problems are best resolved by those who live and work in the watershed (IJC 2005). This belief mirrors trends in environmental decision-making more generally, where it has been recognized that the inclusion of extra-governmental participants has both practical benefits, such as more detailed an intimate knowledge of environmental systems, and normative ones, including the fact that it is more democratic.

Summary of Commission Activities

Phase 1 (1998–2000)

Throughout 1998 and 1999, commissioners consulted federal, provincial, and state officials and held workshops in areas the Commission identified as possible sites for the first international watershed board (Rainy, Red, and St. Croix watersheds). In many cases, the officials being consulted were already members of IJC boards, but they would now be responding in their agency capacities and not in their "personal and professional" capacities as IJC board members. Somewhat surprisingly, while the watershed concept itself met overall support, the reaction of officials and personnel from many government departments and nongovernmental organizations was not supportive of a watershed board in their watershed for various reasons.

Phase 2 (2001–2005)

The Commission's First Progress Report (IJC 2000) summarized activities that had taken place under the 1998 reference and noted some of the issues encountered in consultations with federal, state, and provincial officials. Unfortunately, it led to misunderstanding about whether the IJC was recommending new boards or simply enhanced mandates for existing IJC boards, and further, whether any type of IJC board would assume some responsibilities of other existing agencies. This misunderstanding, among other issues, contributed further to difficulties in implementing international watershed boards. In 2002, in an effort to better explain the concept, IJC staff and newly appointed IJC commissioners began working closely with existing IJC boards and their watershed partners to clarify the vision of the IWI. IJC commissioners, staff, board members, and government officials agreed that incremental advancement toward the full watershed concept, beginning with enhancing the capabilities of several existing boards, would be the most practical strategy for building effective and respectful arrangements. IJC boards would continue their current responsibilities and, in full cooperation with other entities, would slowly and carefully build partnerships to improve local capability in monitoring and addressing transboundary water and related environmental concerns. This modification of the concept the Commission originally proposed in its first progress report became the core idea of the modified international watershed board concept and led to a rebranding of the concept into the current IWI.

Phase 3 (2005–present)

During this period, IJC commissioners and staff worked to strengthen the capacity of the IJC boards and provide funding for selected projects, such as harmonized transboundary watershed maps and standardized hydrographic data, river and reservoir hydraulic models, and expansion of public outreach. For the 2005–2012 period, the two federal governments responded by providing or pledging a total of nearly $4 million (CDN), in roughly equal shares. This funding allowed the IJC to support more than 30 projects in the four pilot watersheds (see Figure 4.1).

Figure 4.1 IJC International Watershed Boards.

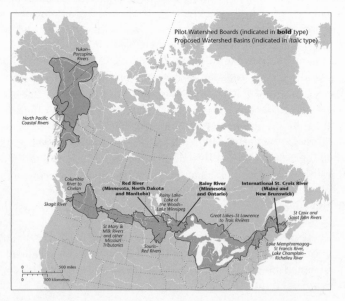

Regardless of all the potential benefits an IJC international watershed board offered – including building capacity at the watershed level, coordinating government institutions, consulting with and involving the full range of interests concerned, and having the flexibility to identify and handle unforeseen developments, all under the umbrella of a permanent treaty-based institution – commissioners initially failed to generate a sufficient level of interest from a wide enough spectrum of entities in any one watershed to galvanize the formation of an international watershed board in any of the watersheds they visited. The Commission reluctantly concluded that, in the absence of more substantial support from federal, provincial, and state governments and provision of the financial and other resources that those boards would require, it was not yet feasible to establish IJC international watershed boards as originally conceived in the 1997 Twenty-first Century Report.

Commissioners decided to make a start and provide many of the basic benefits of IJC international watershed boards by amalgamating IJC boards in watersheds where the Commission already had water quantity and water quality boards: the St. Croix River, Rainy Lake – Rainy River– Lake of the Woods, and the Red River. They also amalgamated the Souris River responsibilities of the International Souris – Red River Engineering Board and the International Souris River Board of Control. With the support of Canadian and US government officials, these amalgamated boards began to adopt a more integrated and ecosystem approach to transboundary issues that fell within their mandates. This ecosystem approach, when combined with the boards' traditional independent, binational structure, and their commitment to finding consensus that reflected the common interests of both countries, enabled the amalgamated boards to begin to function somewhat as watershed boards. Commissioners hoped that, over time and with the addition of nongovernmental members, these amalgamated boards would start to evolve into the international watershed boards the 1997 report envisaged. This incremental approach was an important first step in the evolution of the international watershed board concept and proved to be less threatening to those living in the various transboundary watersheds, addressing the perception that the IJC was "taking over" responsibilities in the various watersheds.

Challenges

Why did these boards not take off as expected? During the phases summarized above, the Commission faced, and continues to face, several challenges along the road of establishing and maintaining international watershed boards. These challenges are not limited to the IWI – as discussed below.

Jurisdictional Challenges

At the federal level, one significant challenge was the uneven response from officials in Canada and the United States. Officials in Canada expressed concern that they could be asked or required to divert scarce resources away from other priorities to support the IWI. Federal agencies in the United States however, were generally more optimistic, and some, such as the Environmental Protection Agency (EPA), were enthusiastic about the benefits international watershed boards offered.

The Congress of Aboriginal Peoples (CAP) expressed support for the concept of international watershed boards. I discuss more broadly here the relationship between First Nations Peoples and the IJC. While it has been made clear that the CAP would like to be involved at the highest level (that is, to have a representative act as an appointed commissioner), it has participated actively in several IJC studies and participated on IJC boards. Due to the wide geographical, cultural, and organizational variation within native submissions and communications, it is difficult to identify a singular narrative. Broadly speaking, IJC areas of Aboriginal interest in the west tend to generally deal with specific projects or activities, while IJC areas of interest in the east have a much larger span of concerns. Indeed, the Great Lakes – St. Lawrence River ecosystem provides the most in-depth view of the concerns of, and involvement by, First Nations in IJC work.

At the provincial-state level, the Commission found less support for the IWI concept. Although there was some variation in their views, British Columbia and Washington, as well as New Brunswick and Maine, considered their existing provincial-state arrangements (such as the Washington – B.C. Environmental Cooperation Council and St. Croix International Waterway Commission) adequate to address most transboundary water issues. The mood at one of the first meetings involving IJC commissioners with officials and others from New Brunswick and Maine was so sombre and unenthusiastic that I thought the whole watershed board concept might have to be shelved. Instead, everyone involved in these consultations agreed to "go slow" rather than abandon the concept and wait until "the timing was better," there was stronger local support, and there was less concern that the IJC was "taking over" water management.

Financial Challenges

Financial resources have frequently been a challenge at the IJC, and the rollout of the IWI proved to be no exception. At the same time as the Commission had been carrying out its consultations on the watershed reference, it had also been completing its work on flooding in the Red River Basin, under a separate reference from governments (Minister of Foreign Affairs 1997). While working on both of these references, the Commission was impressed with the energy various officials in both countries throughout the Red River Basin were devoting. Indeed, in light of the many federal, state, provincial, and local watershed initiatives in that basin, the Commission, in its 2000 report, recommended that governments approve the International Red River Board as a prepilot effort and provide sufficient financial resources to fund international watershed board capabilities for the prepilot project. The Commission estimated that US$165,000 annually would fund this initiative to cover, among other things, ongoing informational networks and outreach activities, intergovernmental liaison, and information gathering and reporting. The Commission anticipated that as each watershed board came into existence, a similar level of funding would be required annually for each board, and commissioners believed that without such additional resources, the international watershed boards concept would not succeed. Governments did not provide these funds initially and, as discussed below, the IJC itself later revised/increased the amounts and purposes for which the funds would be needed; the governments subsequently endorsed the IWI and provided funding.

Following the identification of the three pilot sites for the IWI in its 2005 report (the St. Croix River in New Brunswick, Maine; the Red River, principally in North Dakota, Minnesota, and Manitoba; and the Rainy River in Minnesota, Ontario), the Commission noted the significant funding issue once again and pointed out that moving forward on IWI projects required new and dedicated funding from the two federal governments. The Commission estimated that implementation of all projects would cost US$600,000 per year over a three-year period. Of the three-year total of US$1.8 million, $360,000 would be for core operational funding and $1.44 million would be for specific-project, one-time funding. The US section of the IJC planned to spend several hundred thousand dollars from its 2005 fiscal year operating budget; however, the Canadian section, aside from funding support for secretarial services for the International Red River Board, did not have discretionary funding to match these allocations. This funding asymmetry presented an important challenge, given the Commission's principle of equal cost sharing between the US and Canadian sections. In 2007, the Canadian section was successful in obtaining funding from the Canadian government to enable it to catch up to the funding the US section had already spent. This Canadian funding consisted of smaller but significant amounts from 2007 to 2012 and a long-term larger increase in its base budget from 2012 onward (Government of Canada 2007).

Governance and Management Challenges

Even with an agreed-upon concept and finances in order, the IWI still faced obstacles. Most notably, the proposed IWI structure was different from the typical IJC board structure, a difference that necessitated further negotiation and restructuring.

By mid-2008, the IWI had matured, several IJC boards had undertaken some significant work, and the Commission decided that a new governance and management approach was necessary to enhance transparency and accountability in the submission, evaluation, selection, and implementation of IWI projects. Commission boards themselves were asked to become the prime initiators of project requests that fit within each board's prioritized work plan. It was agreed that project proposals would be screened and evaluated by IJC staff against clearly articulated criteria and the results shared with all boards. Approved projects would be monitored by staff while being implemented and the results evaluated.

The first tranche of 10 proposals submitted by IJC boards covering the Rainy, Red, Souris, St. Croix, and Osoyoos basins and amounting to about CDN$545,000 were approved in September 2008. The successful proposals demonstrated how each board had identified and acted on its own priorities. An analysis of the projects showed that most of the IWI resources went to hydrologic as well as ecosystem data collection and analysis, data harmonization, public outreach, and overall IJC board support.

Data, Information, and Modelling Challenges

Over the years, each country has developed its own geographic information system (GIS) data sets, but these data sets stop at the international border. Such disconnects hampered local efforts to develop an integrated understanding of transboundary basins. The challenge of collecting, harmonizing, and sharing data across the border is not a new one, but funding from the IWI facilitated these critical processes. Under the IWI umbrella, the IJC has supported work to produce consistent and comparable geographic data sets for watersheds along the boundary since 2005, starting with a pilot effort in the St. Croix Basin.

The IJC is also promoting the development of hydrological models – computerized conceptual representations or simulations of the movement and flow of water into, through, and out of watersheds. These models can be useful to local communities, among other purposes, in efforts to: predict flood and drought risk, reduce their vulnerabilities to such events, manage agricultural productivity, predict geomorphologic changes such as erosion or sedimentation, enhance operation of dams, and assess the impacts of natural and anthropogenic environmental change on water resources. Other activities the IWI now supported include: collection of basic hydrological, water-chemistry, and biological data; analysing trends in water quality; developing protocols and methodologies for environmental monitoring; analysing regulatory frameworks and legal regimes governing water resources; and preparing reports and maps to increase public knowledge and awareness of various watershed issues. All or even some of the above activities and projects might have evolved without the introduction of the IWI, but it is doubtful. The IWI provided the needed "spark" of a new idea, increased local involvement in watershed issues, and a needed mechanism to obtain additional funding, without which it would have been difficult for the IJC and its existing non-IWI boards to promote and implement such new and innovative activities.

Case Study 1 – St. Croix: The First Pilot Project

For many years, the IJC had two boards in the St. Croix River watershed, one dealing with water levels and flows and another concerned with water quality. The International St. Croix River Board of Control was responsible for quantity (levels and flows), and the International Advisory Board on Pollution Control was responsible for quality. The two boards had already been working well together for some time and had regularly held joint annual public meetings. In 2000, these two boards merged – the first step in becoming an international watershed board. Yet it took another seven years before everyone (the board members, the IJC commissioners, nongovernmental organizations, government officials at all levels, local interest groups, the public, and the media) was comfortable enough to want the board to have the official title of "international watershed board."

Recent accomplishments of the board under the IWI include: development of a GIS Atlas of the Watershed; merging data from Canada and the United States to develop a seamless, harmonized set of watershed boundaries and watershed stream networks; and preparation of a State of the Watershed Report, which summarizes available information on the basin for a broad audience (IJC 2011). The board also has taken a leadership role in promoting and disseminating scientific information regarding the management of migratory river herring (alewife), thus contributing to the ongoing discussions that may lead to modifications of Maine legislation that has resulted in blocked fishways on the St. Croix River. Future board-supported efforts may include additional data/map products and organization of forums on topics of interest on both sides of the border.

Success in this watershed is likely due in large part to its relatively small size, the somewhat limited number of governments and government agencies involved, an adaptive management approach, an active and involved group of citizens, board members who have similar objectives and who have been able to cooperate effectively, the long history of IJC involvement in the watershed, and realistic expectations of what the IWI and an IJC watershed board can accomplish.

Case Study 2 – Red River: Building on an Ecosystem Approach

The IJC's involvement in the Red River Basin is extensive and began with a January 1948 reference on water use and apportionment. Fifty years later, the two federal governments asked the Commission (through a joint reference in 1997) to identify longer-term solutions to ongoing flooding in the basin. A report in 2000, titled *Living with the Red – A Report to the Governments of Canada and the United States on Reducing Flood Impacts in the Red River Basin,* set out a number of conclusions and recommendations to provide for a more flood-resilient basin. By this time, however, the IJC had already begun moving toward the IWI projects and recommended including in the development of the International Red River Board (IRRB) the alerting and monitoring functions the initial IWI vision had identified .

The current board's mandate is defined by its directive dated 7 February 2001 (IJC 2011). Its activities focus on water quality, water quantity, levels, and aquatic ecological integrity in the Red River. The IRRB is not yet a true international watershed board. Nevertheless, it has adopted an ecosystem approach to its work, and it is hoped that the board will soon attain watershed board status in a manner similar to the St. Croix watershed. It is quite possible that once certain complex relationships in the basin are clarified and "sensitive environmental issues" (including Devils Lake flooding, Pembina road-dike, and Northwest Area Water Supply Project) have been addressed, the Commission's designation of an IJC international watershed board might be accepted. In the interim, the board has been instrumental in addressing environmental issues arising from the diversion of water from Devils Lake into the Red River watershed (chapter 9), working toward the setting of nutrient objectives for the Red River at the international boundary, developing water-quantity apportionment procedures, and overseeing hydraulic modelling efforts to help understand and alleviate episodic flooding problems in the lower Pembina River Basin. Of particular note is the work this board is doing, under the aegis of the IWI, to monitor fish parasites and pathogens in Devils Lake pursuant to a government request (International Red River Board 2007, 2009; see also chapter 9, this volume). This is an excellent example of how the IJC is working on sensitive transboundary issues without a formal reference from governments, and it could be indicative of how the IJC will serve both countries in the future.

Analysis of the IWI

The Commission and its boards have developed an IWI framework and process that recognizes the uniqueness of each watershed and the need for local involvement, the essential elements of which are described in the following section.

Local Involvement and Public Participation

The IWI creates opportunities for exchanges of views, knowledge, and information among all those interested in an issue, which again furthers the development of understanding and consensus. Unique to the IWI, all commissioners try to implement the core idea that local people and institutions are often the best placed to anticipate, prevent, or resolve many problems related to water resources and the environment, and to take shared actions toward shared sustainability objectives. As a result, board members and local citizens work closely together on issues, development of work plans, and board activities. IJC commissioners and board members make a special effort to build a spirit of openness and trust with those who live and work in the basin, visit and attend local meetings as often as they can, and support the combined efforts with IJC resources, including people and funds. This approach is starting to pay real dividends and should provide a firm foundation for transboundary work in the future. A testament to the growth is the increased dialogue concerning the approval of a Lake of the Woods and Rainy River international watershed board. In August 2012 the federal governments approved such a board, and in January 2013 the IJC formally created the International Rainy Lake of the Woods Watershed Board from an amalgamation of the International Rainy Lake Control Board and the International Rainy River Pollution Board, with water quality responsibilities for Lake of the Woods. In April 2013 the IJC issued a directive to this new watershed board.

As discussed in the previous chapter, it is important to remain cautious about conflating increased participation with institutional capacity and decision-making of local actors. However, I realize that through sustained engagement and ongoing dialogue, the conditions are being created for long-term collaboration. Whether this moves from an advisory role to a decision-making role is yet to be determined; but from my experience, I remain optimistic.

Government Support

If, however, the IWI is to continue moving forward, it will require ongoing support – financial, human, and logistical – from all levels of government, including federal, provincial, municipal, and First Nations. Recent responses (DFAIT 2011; US Department of State 2011) from both federal governments to the IJC's 2009 Third Report to Governments indicate a high level of support for the Commission's efforts to date. Of note, the US response has indicated a hope that the Commission will expand consultations and cooperation with tribal governments and First Nations, assisting with strengthening and partnering with such governments in the binational watersheds. It will be interesting to see how the Commission responds to this request in its future projects and reports.

Dedicated IJC Involvement

IJC commissioners and staff have worked hard to strengthen the existing IJC boards, help two of them (St. Croix and Rainy Lake of the Woods) to move from the pilot stage to the full-fledged international watershed board status, and nudge other non-IWI boards closer to the IWI vision. A transparent and accountable management and governance system for awarding and reporting on how IWI funds are spent and what benefits are derived from IWI activities has proven very helpful. The Commission regularly reports to governments and the public on the progress being made and alerts the governments of any issues. The IJC continues to be, as needed, a catalyst in the harmonization of environmental data and information in the transboundary basin by, for example, strengthening its GIS capacity. The Commission is also working with the boards to help them address emerging issues, such as climate change and health effects, by providing expertise and information from major IJC studies being undertaken, and through the IJC's Health Professionals Task Force. The Commission continues to identify and recommend other boards that may benefit from becoming an international watershed board and explores alternative models or mechanisms for implementing a watershed approach in basins where establishing a board is not feasible or necessary. Finally, the Commission is expanding its outreach to provincial, state, and local officials and institutions and nongovernmental organizations, encouraging their participation in the IWI.

Size

One issue that adds to the complexity of achieving international watershed board status is the varying scale/drainage area of the pilot basins. Larger watersheds appear to bring more potential contentious issues to the table, as well as make public involvement and engagement more difficult. What affects some of the citizens in one area may not have any effect on the majority of residents elsewhere in the basin. It appears that in the St. Croix, the relatively "manageable size" of the watershed may be an important factor, as was discussed in the case study above.

Summary and Reflections

After more than a decade of implementing the watershed concept and the ecosystem approach in various Canada-US transboundary watersheds, the IJC has learned much from the IWI's challenges, failures, and successes, and these might have applicability beyond the IWI context. Two lessons in particular are worth highlighting here. First, initial reservations on the part of provincial/state government officials and local nongovernmental agencies highlighted the importance of social and governance issues (including funding), which proved to be just as important as – if not more important than – technical or engineering challenges. Second, the IWI experience emphasizes the importance of incremental change rather than radical overhauling. The IWI came back from the brink of fading out once the Commission softened the language and proposed a step-by-step approach beginning with a restructuring of existing, familiar entities. Related to this, of course, is the oft-repeated mantra that there is no "one size fits all" solution to these complex issues. And perhaps there is some applicability of the IWI and the watersheds board concept in the reconsideration of the Columbia River Treaty, currently underway (see chapter 7, this volume).

One can easily see how these insights can be used to reflect on the IJC as an organization more generally, and how the language of the BWT aligns more smoothly with the incremental approach that is so critical to the flexibility of the IJC. The IWI also speaks to the increasing importance of ecosystem functions and First Nations, which are reflected in the IWI but not explicitly in the BWT. The ability of the Commission to adapt to these changing realities while remaining true to the original treaty is a testament to the flexibility, adaptability, and strength of existing transboundary institutions. As is seen in the flashpoints chapters (especially the chapters on the Columbia River [chapter 7] and Devils Lake [chapter 9]), these new issues pose the greatest challenges to today's transboundary arrangements.

Note

1. For thoughtful analysis of the IJC and its operations, see Schwartz (2006); Fischhendler and Feitelson (2005); and LeMarquand (1993).

References

Canadian Water Resources Association (CWRA). 2004. *Canadian Perspectives on Integrated Water Resources Management*, ed. Dan Shrubsole. Cambridge, ON: Canadian Water Resources Association.

Cohen, Alice, and Seanna Davidson. 2011. "The Watershed Approach: Challenges, Antecedents, and the Transition from Technical Tool to Governance Unit." *Water Alternatives* 4 (1): 1–14.

Department of Foreign Affairs and International Trade Canada (DFAIT). 2011. Government of Canada response letter 27.

Fischhendler, Itay, and Eran Feitelson. 2005. "The Formation and Viability of a Non-Basin: The US-Canada Case." *Geoforum* 36 (6): 792–804. http://dx.doi.org/10.1016/j.geoforum.2005.01.008.

Government of Canada. 2007. Budget documents.

International Joint Commission (IJC). 1997. "The IJC and the 21st Century: Response of the IJC to a Request by the Governments of Canada and the United States for Proposals on How to Best Assist Them to Meet the Environmental Challenges of the 21st Century." http://www.ijc.org/php/publications/html/21ste.htm.

International Joint Commission (IJC). 2000. "Transboundary Watersheds: First Report to the Governments of Canada and the United States." December. http://www.ijc.org/php/publications/pdf/ID1563.pdf.

International Joint Commission (IJC). 2005. "A Discussion Paper on the International Watersheds Initiative: Second Report to the Governments of Canada and the United States." June. http://www.ijc.org/php/publications/pdf/ID1582.pdf.

International Joint Commission (IJC). 2009. *The International Watersheds Initiative: Implementing a New Paradigm for Transboundary Basins*. Third Report to Governments on the International Watersheds Initiative. http://www.ijc.org/php/publications/pdf/ID1627.pdf.

International Joint Commission (IJC). 2011. "Boards." http://www.ijc.org/en/home/main_accueil.htm.

International Joint Commission (IJC) United States and Canada. 1972. Great Lakes Water Quality Agreement.

International Joint Commission (IJC) United States and Canada. 1978. Great Lakes Water Quality Agreement Protocol, 22 November.

International Joint Commission (IJC) United States and Canada. 1987. Great Lakes Water Quality Agreement Protocol, 18 November.

International Red River Board. 2007. "Parasites and Pathogens of Fish from Devils Lake, Sheyenne River, Red River and the Red River Delta." Report for the Fall 2006 Program. http://www.ijc.org/rel/pdf/irrb_Fall_2006_Program_Report.pdf.

International Red River Board. 2009. Presentation on the results of parasites and pathogens monitoring in Devils Lake, Sheyenne River, Red River and Lake Winnipeg, Washington, April 2009.

LeMarquand, David. 1993. "The International Joint Commission and Changing Canada – United States Boundary Relations." *Natural Resources Journal* 33:59–91.

Minister of Foreign Affairs. 1997. IJC reference letter 12 June.

Schwartz, Alan M. 2006. "The Management of Shared Waters." In *Bilateral Ecopolitics: Continuity and Change in Canadian-American Environmental Relations*, ed. Philippe LePrestre and Peter Stoett, 306. Burlington, VT: Ashgate Publishing.

US Department of State. 2011. Government of United States response letter, 7 March.

Water Resources Institute (WRI). 2000. "A Guide to World Resources 2000–2001: People and Ecosystems, The Fraying Web of Life." Washington D.C. World Resources Institute (WRI).

World Meteorological Organization (WMO). 2010. "WMO Guide to Hydrological Practices." WMO-168. Part F: Applications for Water Management.

5 Continental Bulk Water Transfers: Chimera or Real Possibility?

FRÉDÉRIC LASSERRE

North America is a continent blessed with large volumes of freshwater. However, its hydrology is uneven: the Great Lakes are replenished only at an annual rate of 1 per cent, and 60 per cent of Canada's water flows northward. The mostly arid southwest, where the United States' population is growing the fastest, struggles to meet the needs of its urban and rural populations. Yet with North America's watersheds divided by geopolitical boundaries, most water issues are dealt with on a national basis, with international agreements helping to resolve transboundary issues when they arise. This pattern has led to calls for a continentalization of North America's water resources through a networking of its watersheds. Because of these calls, this chapter focuses on the long-standing debate about water exports between Canada and the United States, a debate whose nature is quite different in each country.

This chapter argues that transboundary governance, aside from being highly desirable, is also contentious, especially regarding bulk water exports. The governance of water transfers is generally a matter of substate and state jurisdiction, each one jealously protecting its own water resources and avoiding any long-term commitment aimed at sharing them. In fact, the increased involvement of the civil society, sovereignty issues, and water security makes the transboundary governance of bulk water exports highly improbable on the short and middle term.

In Canada, the public is wary about transboundary bulk-water supply schemes with the United States. Fears surrounding such exports have frequently flared up in political debates over most of the past 60 years, contributing to the kindling of a hydro-nationalism that considers water as a national symbol and a national patrimony deserving special treatment. Canadians are widely reported to be deeply attached to their water resources (Barlow 2007; FLOW 2010), which are seen as akin to a national heritage. However, Canadians make very poor use of those resources due to overconsumption, heavy pollution (with municipal and agricultural wastes), and in many areas, a total reconfiguration of the waterscape that negatively affects the local wildlife (Francoeur 2006). What many Canadians do not realize is that their country is already the largest diverter of water in the world, which has helped to make its hydrological engineering firms among the very best. Major diversion schemes – in which water is transferred from one watershed to another – began in Canada at the beginning of the twentieth century, and more are being built, particularly in Québec.

By contrast, the United States has no comparable water export debate. Clearly, this is because Canada does not constitute a suitable market for US water and because there has never been a plan to export water to Canada. There has been some discussion about importing water from Canada, though the idea has never received the widespread attention the export debate in the latter country generated. Mostly during the 1960s, and more sporadically since then, calls have been made to bring in water from Canada to quench the thirst of the southwestern states (Moss 1967; Wright 1966; Bocking 1972; Holm 1988). While the Canadian public has loudly voiced its opposition to such schemes, the US public has been mostly indifferent to the issue thanks to flattening water consumption, higher water uses, productivity, and effective consumption control. Also, no strong political support exists for diversion schemes. Even the Western States Water Council estimated continental diversion projects were unlikely ever to be built (Tony Willardson, e-mail correspondence with the author, 22 February 2005). In most cases, local water stresses have been resolved locally, with water-transfer schemes taking place mostly within national borders, most frequently within states, and less often between states.

The water export debate has led many Canadians to believe that Americans are actively pursuing a takeover of Canada's water resources. In fact, Canadians, including entrepreneurs, engineers, civil servants, professors, and farmers, designed the majority of continental bulk-water transfer schemes. These proponents were interested in promoting water exports in the hope that Canada, and often that they themselves, would benefit financially. Importantly, the Canadian and US governments did not directly and actively support these schemes –despite some occasional positive comments – nor have they actively sought to bring them to fruition.

This chapter addresses the issue of continental water transfer schemes between Canada and the United States. It seeks to debunk some persistent myths about such water transfers and to answer a question that has frustrated generations of scholars and political figures: are continental bulk water transfers from Canada to the United States a chimera (an illusion), or do they have real potential?

Pipe Dreams: Continental Water Transfers

A continental bulk-water transfer is a transfer of water initiated through human actions; this transfer aims to move water in large quantities from one watershed to another and involves crossing an international border. To do so, an extensive network of water-related infrastructure would have to be built, including canals, diversions, pipelines, dams, pumping stations, hydropower dams, or nuclear plants. The water would be subject to a commercial transaction or to market mechanisms, and once transferred across the international border, would be considered a water export.

As discussed further in chapter 6, the past 60 years have seen many proponents who promote and defend a vast array of schemes (more than half of these creators were Canadians, mostly individuals acting on their own; see Forest and Forest 2012) aimed at transferring water from "have" to "have-not" regions all across North America. Indeed, years before Canada and the United States (and later Mexico) implemented a free-trade zone, these schemes were promoting the continentalization of this natural resource through the pooling and networking of watersheds into a continental water grid. People believed that the relatively plentiful water resources in the north could be drawn upon to benefit the south. The logic of these plans depended on long-term projections (sometimes in the hundreds of years) that themselves were based on population growth rates from the late 1950s and early 1960s, and that predicted catastrophic water shortages that would impede and threaten the economic and demographic development of the American West (Forest and Forest 2012). Some politicians supported these schemes, at least partly because the schemes would have generated large investments in their ridings. They thus called for the watering of the American West in response to "[t]he coming water famine" (Congressman Jim Wright [Texas, D.], 1966) or "[t]he water crisis" (Senator Frank E. Moss [Utah, D.], 1967). Senator Moss was the most outspoken supporter of the schemes, notably the North American Water and Power Association (NAWAPA), which called for the diversion of Arctic rivers to the southwest.

A wide range of continental water schemes emerged (Table 5.1) from the end of the 1950s on, all aimed at balancing the perceived uneven continental distribution of water, but also at bridging the gap between the growing demand and water availability. The plentiful but underdeveloped water resources of the north were targeted as the source for the future water supplies. With the help of large-scale engineering projects, a whole range of infrastructures would have had to be built to supply the southwest with the resource that was scarce there, namely water. These would have included dams as tall as the Empire State Building, pumping stations requiring massive amounts of energy – equivalent to that being produced by Québec's James Bay development – as well as tunnels, diversions, hydroelectric dams, or nuclear plants.

Most proponents of such schemes estimate price tags in the billions of dollars, but still manage to severely under evaluate the prices by neglecting the

Table 5.1 Water Export Proposals from Canada

Project	Year	Source	m³/s
GRAND Canal	1959	James Bay watershed diversion toward the Great Lakes and the Western United States	3 361 to 11 000
Great Lakes Transfer Project	1963	Skeena, Nechako and Fraser in British Columbia; Athabasca and Saskatchewan in the Prairies, all towards the Great Lakes	4 501
North American Water & Power Alliance (NAWAPA)	1964	Transfer from Pacific and Arctic watersheds to the Great Lakes, the Mississippi and the southwest	9 827
Kuiper Plan	1966	Peace River, Athabasca, North Saskatchewan, Nelson and Churchill	5 865
Magnum Plan	1966	Peace River, Athabasca and North Saskatchewan within Alberta	983
Stabilization of the Great Lakes	1966	Hudson Bay Watershed	555 to 1 132
Central North American Water Project (CeNAWP)	1967	Mackenzie, Peace River, Athabasca, North Saskatchewan, Nelson and Churchill	5 865
Western State Water Augmentation	1968	Liard and Mackenzie	1 553
NAWAPA-MUSCHEC (Mexican United States Commission for Hydroelectricity)	1968	Sources for NAWAPA, plus lower Mississippi and Sierra Madre rivers	11 222
North American Waters	1968	Yukon and Mackenzie, Hudson's Bay Watershed	58 646
Water for survival	1968	Arctic Rivers	Unknown
Alaska Subsea Pipeline	1991	Alaska's Pacific Watershed	155
Multinational Water & Power Inc.	1992	Pacific Rivers in British Columbia	42

(Continued)

Table 5.1 *(Continued)*

Project	Year	Source	m³/s
Klymchuk	2001	Manitoban Hudson Bay Watershed	155
Gingras	2009	James Bay watershed derivation in the Ottawa River	793

Sources: Lasserre (2005a); Gingras (2009); Klymchuk (2001); Western States Water Council (1969).

costs associated with environmental and Indigenous issues (see chapter 2, this volume), as well as the legal liabilities. Just as importantly, no economic impact studies have yet proven the profitability of such large-scale schemes. Water transfers over long distances are costly both because they require large initial investments and because the consumers rarely entirely pay for the water transferred. In fact, other than urban dwellers, farmers would be the principal recipients of the water, and with their current water consumption already heavily subsidized, it is hard to imagine that the costs would be recouped. Ultimately, only one fundamental question really matters: does the United States need to import freshwater to satisfy its needs? So far, the response has been an unequivocal no for the reasons discussed later in this chapter.

Canada's Water Export Debate: Between Potential and Hydro-nationalism

Canadian water resources have been celebrated in Canada as a source of national pride, paradoxically at the same time they are heavily impacted by decades of pollution and overuse (Bakker 2007; Sproule-Jones et al. 2008). This disregard is partly explained by the tendency in the national psyche to regard water resources as infinite, plentiful, and renewable, and as an unlimited source of power production (Lasserre 2002, 18–20), none of which remain true. As highlighted by Sprague (2007), the availability of Canada's water resources has generated its own debate thanks to overly generous assessments of the country's available water; some credit Canada with having up to a quarter of the world's water supplies. Yet in fact:

> Canada's annual renewable volume of fresh water seems large, at about 2,850 cubic kilometres, but that is only 6.5 percent of the world total of almost 43,800 cubic kilometres. (Sprague 2007, 25)

Still, with such massive water resources in its territory, and given its willingness to exploit them, it is not surprising that Canada has become a world leader in water diversion schemes and hydrological engineering. In fact, Canada is a much larger diverter of water than the United States, despite having a population that is only a tenth of the US size: its diversions exceed 3,600 m³/s, as opposed to about 990 m³/s in the United States. To give perspective on these numbers, Niagara Falls has an average flow of 5,830 m³/s. That being said, the distances involved in US diversions are much greater: most US diversions easily exceed 250 km, while Canadian diversions are nearly all less than 40 km (with only two exceeding 150 km [Table 5.2]). Moreover, the Canadian diversions tend to be over short distances for hydroelectricity-generation purposes, rather than for consumption; indeed, 97 per cent of all water diverted in Canada is used for power generation (Quinn et al. 2004; Lasserre 2006).

It is important to underline the fact that all the diversions discussed above take place within the limits of a single province (in Canada, see Table 5.2) or state (in the United States, see Table 5.3). There are no interprovincial or interstate diversions: all the projects involving two or more states have been consistently opposed by the source states.

Bulk water withdrawals do not occur without impacts on habitats in both the source and the augmented watersheds. Yet the scale of these impacts remains to be determined, since the concept of minimum ecological flow is still subject to debate among biologists (Postel and Richter 2003; Schroeter et al. 2005). The international academic literature reports that impacts begin when between 2 per cent and 10 per cent of the river flow is being diverted (Nilsson et al. 2005). Although the environmental impacts of smaller diversions may be minimal, the sheer magnitude of some diversions leaves no room for doubt about their environmental impacts. In northern Québec for example, the once mighty Eastmain River is now reduced to a shadow of its former self due to the diversion of 90 per cent of its flow into the La Grande River for hydropower purposes (Hydro-Québec 2001), while 71 per cent of the Rupert and 40 per cent of the Caniapiscau's flow are similarly diverted.

Despite these issues, numerous Canadian politicians or senior government officials have not necessarily been opposed to large-scale water schemes. In 1984, British Columbia Environment Minister Anthony Brummett predicted that within 15 years, a worldwide water crisis would create an export market for the province to sell part of the "overabundance" of water in BC (Keating 1986, 163). Canadian Prime Minister Mulroney (Solomon 1999; Bakenova 2008, 287), Québec Premier Bourassa (1985), and Simon Reisman, chief negotiator for Canada during the free-trade negotiations with the United States (Holm 1988, 34), were reportedly in favour of the 1985 GRAND Canal scheme[1] to build a dike at James Bay and transport freshwater toward the south via the

Table 5.2 Major Existing Large-scale Water Transfer Schemes in Canada, 2009

Scheme	From (basin)	Destination	Beginning of Operation	Transfer Volume (m³/s)	Length of Diversion (km)	Objective
Kemano	Nechako R. (Fraser)	Kemano River, BC	1954	115	18	Hydropower
Coquitlam-Buntzen Hydroproject	Coquitlam (Fraser)	Buntzen Lake and Indian Arm (Burret Inlet), BC	1912	28	4	Hydropower
Vernon Irrigation District	Duteau Creek (Fraser)	Vernon Creek (Columbia), BC	1907	0.6	~5	Irrigation
Gardner Diversion	South Saskatchewan River	Last Mountain Lake, Qu'Appelle R.	1976	3,5	137	Multipurpose, mainly irrigation
Qu'Appelle Diversion	South Saskatchewan River	Qu'Appelle River	1967	2,5	~5	Multipurpose
Churchill	Churchill	Nelson, MB	1976	775	~40	Hydropower
City of Winnipeg Aqueduct	Lake Shoal (Great Lakes)	Winnipeg (Red River), MB	1919	3	137	Urban
Great Lakes Basin	Long Lake	Lake Superior	1939	42	0.4	Hydropower
Great Lakes Basin	R. Okogi (Albany River Basin)	Lake Superior	1943	113	8.5	Hydropower

Root River	Lake St.-Joseph (Albany River)	Root River (Nelson), ON	1957	86	7	Hydropower
James Bay–La Grande	Caniapiscau River	La Grande River, QC	1985	795	250	Hydropower
James Bay–La Grande	Eastmain River	La Grande River, QC	1985	835	150	Hydropower
Manouane	Manouane R. (Saguenay)	Péribonka R., QC (Saguenay)	1961	116	2	Hydropower
Manouane 2	Manouane R. (Saguenay)	Betsiamites R., QC	2003	30,3	7	Hydropower
Sault-aux-Cochons	Sault-aux-Cochons R.	Betsiamites R., QC	2003	6,5	~2	Hydropower
Portneuf	Portneuf R.	Betsiamites R., QC	2003	10,9	?	Hydropower
Cabonga-Dozois	Gatineau R. (Ottawa)	Ottawa R., QC	1928	11,3	~2	Hydropower
Bay d'Espoir Hydropower	Bear, Victoria, Salmon Rivers	Bay d'Espoir, NF	1970	185	200	Hydropower
Churchill Falls	Jultan R.	Churchill, NF	1971	196	?	Hydropower
Churchill Falls	Naskaupi R.	Churchill, NF	1971	200	20	Hydropower
Churchill Falls	Kanairktok R.	Churchill, NF	1971	130	25	Hydropower

Sources: From several sources compiled by the authors, including Day and Quinn (1992); Hydro-Québec (2001); Lasserre (2005b); Saskatchewan Watershed Authority, e-mail correspondence with F. Lasserre, 28 September 2011.

Table 5.3 Major Existing Large-scale Water Transfer Schemes in the United States, 2010

Scheme	From (basin)	Destination	Beginning of Operation	Transfer Volume (m3/s)	Length of Diversion (km)	Objective
Croton Aqueduct	Croton (Hudson)	City of New York	1842	14.2	66.4	Urban
Catskill Aqueduct	Catskill (Hudson)	City of New York	1915	24.4	147	Urban
Delaware Aqueduct	Delaware	City of New York	1952	40.3	169	Urban
Chicago Diversion	Lac Michigan	Mississippi	1848, 1900	91	~40	Navigation, urban, irrigation
Southwest Pipeline Project	Missouri (North Dakota)	Southwestern ND	2010	0.2	~300	Urban, industrial
Canadian River Project	Canadian R. (Arkansas and Mississippi)	Cities of Lubbock, Pampa, TX	1968	2.4	515	Urban, industrial
Los Angeles Aqueduct	Owens River	City of Los Angeles	1913	14	373	Urban, hydropower
Second Los Angeles Aqueduct	Owens River and aquifer	City of Los Angeles	1970	8.2	219	Urban
Hetch Hetchy Water Supply	Tuolumne River	City of San Francisco	1934	11.4	240	Urban
"Continental Divide" diversions	Upper Colorado Basin	South Platte and Arkansas (Missouri)	As of 1892	18.8 overall		Irrigation

Colorado River Aqueduct	Colorado	Metropolitan Water District of Southern California	1941	47.4	387	Urban
All American Canal	Colorado	Imperial Irrigation District	1942	121.4	325	Irrigation
San Juan Chama Project	San Juan River (Colorado)	Chama River (Rio Grande)	1970	4.3	13.8	Urban, industrial
Southern Nevada Water Project	Colorado (Lake Mead)	City of Las Vegas	1971	26.4	35	Urban
Central Utah Project (CUP)	Colorado	Utah	2004	10.6	242	Irrigation, urban
Central Arizona Project (CAP)	Colorado	Arizona (Tucson)	1993	58.7	528	Irrigation, urban
Central Valley Project	Trinity, American, San Joaquin, Sacramento	Central California	1951	273.9	~600	Irrigation
California Aqueduct, State Water Project (SWP)	Sacramento River	Central and Southern California	2003	199.8	710	Irrigation, urban
Rio Colorado-Tijuana Aqueduct	Colorado	City of Tijuana	1985	3.2	150	Urban
Santa Ynez Development Project	Santa Ynez River	City of Santa Barbara	1956	1.01	42	Irrigation, urban

Sources: From several sources compiled by the authors, including Humlum (1969), Day and Quinn (1992), Lasserre (2005a, 2005c).

Great Lakes. According to Bocking (1972, 32–3), the 1960s showed a wide range of Canadian politicians supporting, or at least not closing the door to bulk-water exports; they included Minister of Energy, Mines, and Resources Joe Greene, Prime Minister Lester Pearson, Minister of the Environment Jack Davis, and Jean Chrétien, with the latter affirming in an interview to *Time* magazine on 2 May 1969 that "within 25 years we will be exporting water" (Bocking 1972, 33). In 2001, Newfoundland and Labrador Premier Roger Grimes was initially favourable to the Gisborne Lake export project promoted by Canadian entrepreneur Gerry White, before it was cancelled at last in 2002. Thus, for years, the Canadian government's position seemed to incite fears in the population that not enough was done to protect Canada's water resources.

Moreover, the GRAND Canal project is still advocated by its designer, Canadian engineer Tom Kierans. The Fraser Institute published a report in 2010 advocating a calmer and more reasoned look at the potential for water export from Canada (Katz 2010), and the Montréal Economic Institute (MEI) has recently come out with two water-export proposals. Marcel Boyer, an MEI economist, released a paper in August 2008 in which he waxed eloquently, if inaccurately, about the province's reported "water oversupply," although he failed to provide evidence of any real US market for it and didn't discuss the costs of such exports (Boyer 2008). In July 2009, Pierre Gingras, a retired engineer from Hydro-Québec, published his diversion scheme under the auspices of the MEI (Gingras 2009, 2010). It envisioned a project aimed at transferring water from Québec's James Bay rivers into the Ottawa River. These two documents have generated no comment from either provincial governments or their federal counterpart, a sign that such schemes currently have little momentum. See Figures 5.1a and 5.1b for main water diversions in North America.

The US Water Import Debate: A Nonexistent Discussion

Water is a key ingredient in the fabric of Western American society, as environmental historian Donald Worster (1985) has eloquently documented. Twentieth-century American society, powered by the industrial age, decided to harness rivers and aquifers on a large scale. Technological innovations such as nuclear energy were being developed at fast rates, and it was thought that the scaling up of then-existing technologies would be sufficient to address the nation's biggest challenges. At the time, the environment was very much a secondary concern relative to economic expansion. In the 1960s, fears of water exhaustion and ensuing economic catastrophe spurred politicians to consider large-scale water diversions from the Columbia and the Mississippi. The forceful rejection of water export proposals by the Columbia and Mississippi Basin

Figure 5.1a and 5.1b Main water diversions in North America.

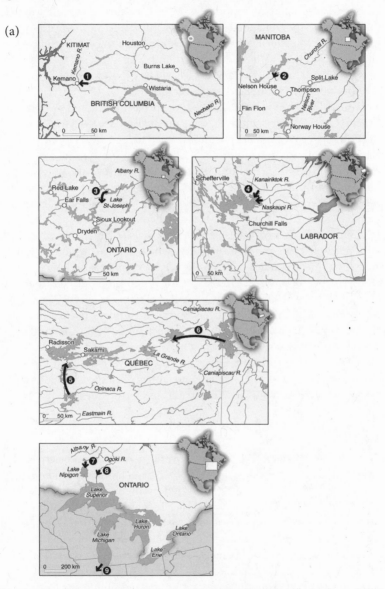

Sources: The Albuquerque Journal, June 25, 2003; International Joint Commission, www.ijc.org ; Hydro-Québec (2001); Lake of the Woods Control Board, *Winnipeg River Drainage Basin,* Winnipeg, www.lwcb.ca/, December 2000; California Water Department; Humlum (1969); SGE Acres, "Hydroelectric Plants and Drainage Areas," *Island Hydrology Review,* Saint-John's 2003, 2004; Lasserre (2005b).

Figure 5.1a and 5.1b (*Continued*)

(b)

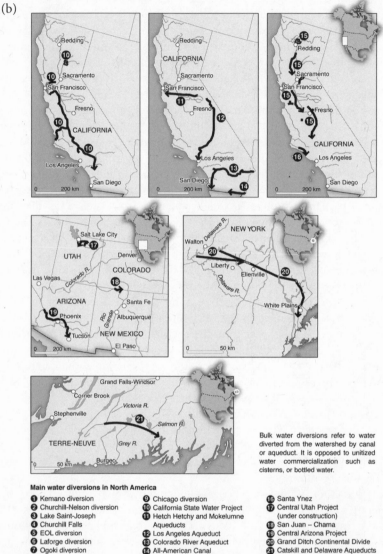

Bulk water diversions refer to water diverted from the watershed by canal or aqueduct. It is opposed to unitized water commercialization such as cisterns, or bottled water.

Main water diversions in North America

1 Kemano diversion
2 Churchill-Nelson diversion
3 Lake Saint-Joseph
4 Churchill Falls
5 EOL diversion
6 Laforge diversion
7 Ogoki diversion
8 Lake Long diversion

9 Chicago diversion
10 California State Water Project
11 Hetch Hetchy and Mokelumne Aqueducts
12 Los Angeles Aqueduct
13 Colorado River Aqueduct
14 All-American Canal
15 Central Valley Project

16 Santa Ynez
17 Central Utah Project (under construction)
18 San Juan – Chama
19 Central Arizona Project
20 Grand Ditch Continental Divide
21 Catskill and Delaware Aqueducts
22 Bay d'Espoir Development

states prompted some to consider the much greater potential available in Canada, and some engineers and engineering firms started designing huge transfer schemes aimed at importing large volumes of water from outside the United States. Under these schemes, Americas' northern neighbour was to be the supply source that would help to address allegedly existing and future droughts.

These draft schemes garnered considerable interest, especially the North American Water and Power Alliance (NAWAPA), which promoted the southward diversion of both Canadian and American Arctic rivers into an enormous reservoir in the Rocky Mountains Trench, and their subsequent distribution as far south as Mexico. A Special Subcommittee on Western Water Development of the Committee on Public Works (1964) studied the NAWAPA proposal and suggested that further studies be undertaken to assess its feasibility. But this interest quickly faded, and in fact, such schemes never entered into the mainstream political debate, with the general population remaining mostly unaware of them. Apart from proposals to take water from Canada, water-diversion proposals were also developed from the American West Coast Basins, like the Columbia, and these schemes usually faced so much opposition that transfers from Canada appeared politically easier to achieve (Lasserre 2005c). These diversion projects generated significant controversy and opposition from the northwest states, some of whom banned or strongly opposed interstate water exports such as the diversion project from the Columbia River (Bocking 1972). Similar opposition emerged in the Mississippi Basin (Bocking 1972; Lasserre 2005c) and in the Great Lakes from the riparian states (Consulate General of Canada 1982; Lasserre 2005d), whereas in Alaska, they were usually favoured (Lasserre 2005b). Alaska boasts large volumes of renewable water, and today, is the sole jurisdiction in North America that is currently open to bids on its freshwater resources. But despite more than a decade of trying to get into the water-export business, Alaska has yet to attract its first customer (Quinn 2007, 7). In 2011, the Southern Nevada Water Authority proposed diverting floodplain waters from the Upper Mississippi to alleviate strain on the Colorado water system (Rowe 2011), a proposal as quickly rebutted by the Mississippi watershed states as the previous ones.

In fact, despite Canadian worries about international water exports (see chapter 6), none of the continental bulk-water transfers have come to pass. The long-term projections from the 1950s and 1960s that emphasised a dire future in which America would experience drought, as well as resultant economic and social disruptions, have proven to be wrong. And while the country's GDP has grown at an exponential rate since then, water consumption rates started to flatten out around 1975 and have remained flat ever since (Gleick 2001; Figure 5.2). This has caused the calls for the continentalization of water supplies and

Figure 5.2 Total water withdrawals in the United States, 1950–2005.

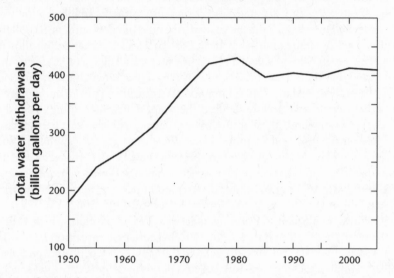

Sources: Original Graph by Eric Leinberger, based on data from Gleick (2010); USGS (1998, 2004, 2009).

distribution to fade slowly into the background, although predictions of a calamitous future still occasionally emerge.

Several factors may explain this disaffection for the Canadian option. First, Americans developed national solutions to address potential water shortages, notably by promoting conservation measures such as xeriscaping in the southwest and building water diversion schemes within intra-state boundaries (Table 5.3). Second, agricultural water usage in the United States peaked in 1980 and then shrank by 14.7 per cent between 1980 and 2005 (USGS, 1998; 2004; 2009). Water pricing and water costs, although still very low thanks to generous subsidies, are gradually increasing (Cline 2003; Artiola and Dubois 1995; Hanak et al. 2009). Generally speaking, whether for agricultural, industrial, or urban use, there is a definite shift away from a supply-side management approach (increasing the volume of available water) and toward a demand-side management approach (more efficient usage of a limited volume of water).

Third, water-starved cities are much richer than the agricultural sector and can afford to buy water rights, build further aqueducts, or develop swap programs. For instance, Las Vegas has a project still pending in 2012; it is being strongly pushed by the Southern Nevada Water Agency (SNWA): the diversion of groundwater from Clark, Lincoln, and White Pine counties in eastern Nevada. The SNWA and the state of Arizona also strongly advocated for swap agreements. For instance, in the case of Arizona, the government has offered to pay for a desalinization plant near Los Angeles in exchange for allowing Arizona to keep the equivalent amount of water from the Colorado for its own use (*Las Vegas Sun* 2005, 2006, 2008; SNWA 2005; Robert Barrett, interview with the author, 9 March 2005, Phoenix). Water transfers from agriculture to the urban sector have greatly reduced the water deficit by helping to satisfy a part of urban demand. These transfers have been possible at least partly thanks to increasing cultivation of more water-efficient crops. An example of such a transfer is the agreement between the San Diego Water Authority and the Imperial Irrigation District, under which irrigators agreed to transfer up to 200,000 acre-feet (about 247 million m³) of water per year for up to 75 years to the city of San Diego (San Diego County Water Authority 2002).

Fourthly, US agricultural producers face fierce competition, which is driving some of them out of business. In 2001, because of competition from other regions (mainly Asia and Mexico), sales of California cotton, prunes, and pistachios fell by about 33 per cent; broccoli and plums by 24 per cent; and tomatoes and lettuce by 22 per cent. As a result, "Californian farmers could soon realize there is more money to make selling their water rights than using them to grow crops" (*The Economist* 2003, 11). Cost incentives also lure US producers to Mexico, further reducing water demand in the United States. As a whole, water withdrawals in the United States have mostly stagnated in the last decades and could even begin a downward trend should competition from foreign fruit and vegetable producers increase.

Thus, even though water is still used at an unsustainable rate in most of the western parts of the United States, importing water from Canada is not as urgent as it once appeared to some Canadian entrepreneurs or US politicians, such as former Senator Paul Simon (1998) (D, Illinois), who actively pleaded for such schemes (Thompson 1999). A stabilizing trend in water withdrawals (Figure 5.2), the availability of cheaper water sources, and water conservation have led public planners to set aside bulk-water diversion schemes from Canada or Alaska (Matheson 1984; Tony Willardson, e-mail correspondence with the author, 22 February 2005).

Great Lakes: the Focus of New Diversion Schemes?[2]

Bulk-water diversions[3] already take place in the Great Lakes, but the balance of incoming and out-of-basin diversions is largely a positive one, with inflow diversions greater than outflow. An average annual flow of about 66 m³/s is diverted into the Great Lakes Basin, overwhelmingly from Ontario diversions (Figure 5.3). The largest out-of-basin diversion is in Chicago. This diversion was built for sanitary reasons in 1900; it allowed water to flow from the Great Lakes into the Mississippi River by reversing the flow of the Chicago River. Since then, the Army Corps of Engineers has proposed increases in the flow of the Chicago diversion and massive new diversions from the Great Lakes Basin, and these have been implicitly supported by the Bureau of Reclamation, especially since acting on these proposals could help to foster the diversion plan from the Mississippi to Texas that the Bureau studied until 1973. Both agencies dedicate most of their energy to promoting hydraulic works and thus were initially favourable to massive water schemes. As mentioned, these proposals have been systematically opposed by riparian states (Consulate General of Canada in Chicago 1982; Lasserre 2001, 2005b). For example, one proposal sought to divert water from Lake Superior or Lake Michigan to replenish the Ogallala Aquifer, but in 1982, the Corps recommended against pursuing this plan because of its major adverse environmental effects and its weak economic underpinnings (Micklin 1985, 54;

Figure 5.3 Major water diversions in the Great Lakes Basin.

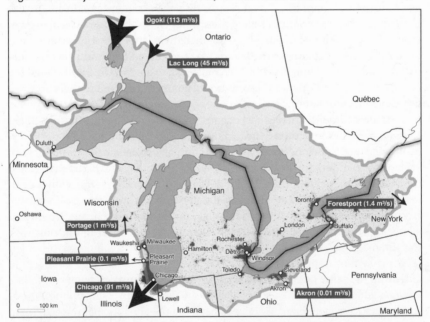

Source: International joint Commission, Protection Of The Waters Of The Great Lakes. Final Report to the Governments of Canada and the United State, Washington, DC, 2000; Lasserre, 2005c.

Keating 1986, 164; Wright and Bulkley 2002, 8). Great Lakes states have traditionally been opposed to these schemes, both for environmental and political reasons. Regarding the political reasons, the logic was to wonder why the Great Lakes states should give water to California at a time when many firms were leaving the area and relocating to the West Coast (Lasserre 2001).

The Council of Great Lakes Governors (CGLG) has been central to the debate over exports from the Great Lakes region (Francoeur 2005). The council, created in 1983, is a partnership of the governors of the eight Great Lakes states, which were subsequently joined by the two Canadian provinces of Ontario and Québec. One of the key results of this transboundary partnership was the signing of the Great Lakes Charter in 1985. That charter, which grew out of a concern that Great Lakes water could be diverted from its watershed, created a notice and consultation process for Great Lakes diversions. The signatories agreed that no Great Lakes state or province would proceed with any new or increased diversion or consumptive use of Great Lakes water exceeding five million gallons per day (19 million litres per day) without notifying, consulting, and seeking the consent of all affected Great Lakes states and provinces.

The riparian states and provinces believed that this charter would prove sufficient to prevent unilateral water export schemes by any signatories, but in 1998, the Ontario government granted an export license to Nova Group to ship water from Sault Ste. Marie to Asia. Although the permit was soon revoked following a public uproar in both Canada and the United States, the fact that the charter had been unable to prevent such a water-export scheme underlined the need for clearer and more binding mechanisms to prevent water diversions. At the same time, public concern in Canada about potential bulk-water diversions was heightened by the persistent, but so far unproven, rumour that President George W. Bush told Prime Minister Jean Chrétien during their July 2001 G8 meeting in Genoa that he wanted the two countries to consider a common pool of natural resources, including energy, wood, minerals, and water (Lee 2001; US Water News Online 2004; Maich 2005). The response to this rumour – a torrent of comments and reactions in the press – illustrates how wary Canadians remain about their water resources, as confirmed in the survey conducted by Nanos (2009), which determined that water is the most vital resource for Canada (61.6 per cent of respondents) and water exports should be forbidden (20.2 per cent of respondents).

During the same period, the issue of supplying water to growing communities located just outside the Great Lakes Basin became a pressing political problem for state governments. More specifically, some constituents wanted to tap the Great Lakes water that lay only a few kilometres away from them through a diversion out of the Great Lakes watershed, while the governors were bound by their pledge not to divert water out of the basin. For example, in 1992, Michigan Governor John Engler vetoed the request from Lowell, Indiana, to access Lake Michigan water, despite the fact that the city's wells were now unsuitable for drinking.

Of course, such water-diversion schemes are radically different from the large-scale, continental schemes envisioned in the 1960s. They are indicative of the trend among recent schemes toward smaller, shorter diversions, which are more politically acceptable and are perfectly justifiable from a good neighbour standpoint. Yet taken together, they could still eventually amount to a large water diversion and could produce severe impacts on the Great Lakes watershed (Lasserre 2005b). Ultimately, the very controversial nature of such diversions has contributed to their non-materialization.

As of 1999, the Canadian federal government has shifted from considering laws that would directly ban water exports to promoting a three-point strategy to deal with the diversion issue. The first step involved a joint commitment with the United States "to study the effects of water consumption, diversion and removal, including for export, from the Great Lakes" (Johansen 2010, 1) under the auspices of the International Joint Commission. The IJC (2000) finally concluded that riparian states and provinces should not allow water diversions out of the Great Lakes.

The second step was to modify the International Boundary Waters Treaty Act to create a regime of restrictive licenses for water removals and to give more powers to the Minister of Foreign Affairs in issuing, renewing, amending, suspending, and revoking licenses for water-removal purposes.

The third and last step was to try to convince the provinces to pass legislation banning massive water removals, since constitutionally, water is a shared responsibility of both the federal government and the provinces. While the federal initiative was rejected by four provincial governments (British Columbia, Alberta, Manitoba, and Québec), all provinces but New Brunswick had independently adopted legislation against bulk water removals by 2003. The latter has still not adopted such legislation (Johansen 2010, 6).

Water Exports and the North American Free Trade Agreement (NAFTA)

The federal government's strategy has been complicated by the trade-related dimension of water exports. That is, the Canadian government's room for manoeuvring is particularly narrow since NAFTA has very stringent rules regarding trade barriers and investor rights. Once bulk water is commodified and traded, it might become very difficult for the Canadian government to limit its export, and once it has authorized Canadian companies to take part in this trade, it cannot prevent US or Mexican companies from also doing so. On the other hand, an explicit bulk-water export ban could be challenged under NAFTA's dispute settlement mechanism as a barrier to trade: if water was recognized as a commodity, Canada could not reduce its exports with nonmarket mechanisms (art. 315), and article XI of the GATT (now part of the WTO Charter) explicitly prohibits bans on exports. True, Canada, the United States, and Mexico signed an agreement in 1993, which states that "unless water in any form has entered into commerce and become a good or product, it is not covered by the provisions of any trade agreement, including the NAFTA." However, this agreement was never ratified and is only a political declaration, and thus has very limited legal value in the face of NAFTA, a ratified treaty.

However, these worries about Canada's capacity to retain its sovereignty over its water resources are counterbalanced by the fact that water is a highly political and contentious issue that NAFTA alone cannot resolve. Moreover, existing provincial legislation and policies would severely restrict or stop any scheme without the consent of political authorities. These laws may of course be repealed, but they do act as barriers to water-export schemes.

Besides, Canadian federal government officials thought that an alternative strategy was possible. This was because the geography of river basins along the Canada-US border meant that most large basins fell neatly onto one side or the other, except for the Great Lakes and Columbia River Basins (indeed, this was a major consideration of negotiators when they settled the international boundary at the forty-ninth parallel; see Nicholson 1979 and Lasserre 1998). In addition, most major drainage basins – from the Fraser Basin to all the rivers draining into the Arctic Ocean and Hudson's Bay – lie wholly in Canadian territory. Thus, provincial governments forbidding out-of-province water transfers was roughly equivalent to preventing water export. At the federal level, forbidding inter-basin transfers on environmental grounds was tantamount to forbidding water exports to the United States without formulating the policy in terms that could make it liable to be challenged under NAFTA. This proved an attractive strategy, since laws forbidding inter-basin transfers could be passed under the guise of environmental protection rather than being written as trade legislation that explicitly banned water exports.

Accordingly, Canada's House of Commons passed Bill C-6, *An Act Amending the International Boundary Waters Treaty Act*, in December 2002. The new law effectively banned bulk-water removals from the boundary water Basins if they were greater than 50 m³/day. Choosing to centre Bill C-6 on watersheds also contributed to efforts to increase protection of the Great Lakes Basin, a body of water that was the focus of the 1909 Boundary Waters Treaty (BWT) between Canada and the United States. With the passing of the amendment to the Boundary Waters Treaty Act, protection for the Great Lakes seems to have been strengthened.[4]

Similarly, in the United States, efforts have been made to prevent large-scale water diversions from the Great Lakes. This underlines the fact that bulk diversions, for now, are advocated only by a minority of lobbyists. In 2000 the Government of the United States (2000) amended article 1962d-20 of the *Water Resources Development Act* (WRDA), so that it now prohibits:

> b.3. [...] any diversion of Great Lakes water by any State, Federal agency, or private entity for use outside the Great Lakes Basin unless such diversion is approved by the Governor of each of the Great Lakes States.

This section gives a veto to any of the Great Lakes governors, thus making any new diversion outside of the basin difficult to authorize without unanimous agreement. Here again, diversions were considered at the basin level: the common unit the act highlighted was the Great Lakes Basin.

The Council of Great Lakes Governors (CGLG, which includes the eight riparian states, Ontario, and Québec), has undertaken a parallel track of negotiations triggered by the 1998 Nova Group export project and the Waukesha import proposal.[5] The CGLG outlined a series of principles for reviewing water withdrawals from the Great Lakes Basin in the Great Lakes Charter Annex of June 2001. The annex showed a real political commitment to thwarting water-export schemes. In December 2005, the Annex Implementing Agreement was signed (CGLG 2005). It made water exports or diversions from the Great Lakes Basin extremely difficult to achieve, thereby alleviating most concerns expressed by Canadians, especially by those in Québec or Ontario. Just like the Canadian and US federal governments, the CGLG decided to focus its response to potential excessive diversions on the Great Lakes watershed, underlining the understanding that any uses at one end of the basin would have effects downstream on other lakes and on the St. Lawrence River. In addition, all states and provinces would use a consistent standard to review proposed uses of Great Lakes water. US states furthermore decided to agree on a Great Lakes – St. Lawrence River Basin Water Resources Compact, signed into law by President Bush in October 2008; the law seeks to ban water diversions.

Future Diversions: A Chimera?

Anti-water export groups are still highly influential in Canadian politics. As recently as November 2008, Prime Minister Stephen Harper pledged in his throne speech to ban bulk-water exports to preserve Canada's environment. Bill C-26, introduced on 13 May 2010, was meant to amend the International Boundary Waters Treaty Act by extending the prohibition on removing bulk water to also cover transboundary waters (water flowing across the border) rather than just boundary waters. Although Bill C-26 died on the order paper because an election was called, a new bill seems likely given that the original bill was introduced by the Conservatives, who have since won a majority government in May 2011. This is the last in a long series of actions taken to protect Canadian sovereignty regarding the country's natural resources. Given the current dissatisfaction some civil society organizations have expressed, more actions can be expected in the upcoming years, especially since strong criticism has been expressed among academics as well, underlining that if Bill C-26 would indeed regulate diversions from transboundary waters, it still said nothing about watersheds farther up north, which therefore in theory can be diverted (Pentland and Hurley 2010).

While large-scale schemes are not likely to happen any time soon due to their complexity cost (see Table 5.4) and the legal Pandora's box they risk opening, smaller-scale water transfers are possible. Some water transfers between borderland communities along the Canadian – US border have already been in place for more than a century without attracting too much attention or opposition from civil society or the media (Forest 2010a, 2010b). These were created as common transboundary responses to local water issues, and they do not involve any water commodification since water is shared as a public service between local public utilities. In themselves, they illustrate the fact that limited and local water transfers can be successfully implemented. They also highlight the multiscalar processes at work with locals engaging the international realm to find concrete and flexible solutions to their problems while enforcing national and provincial/state legislation. As such, these scale entanglements are reflective of good neighbourhood behaviour aimed at cooperating for the common good. The 1990s also saw a multiplication of proposals for another type of water transfer: commercial water export schemes by bulk carriers, designed by private companies to supply communities along the US West Coast or other communities worldwide. Due to public opposition, they have never materialized, but their revival cannot be discounted and depends on the economics of freshwater shipping and the promises of new job creation.

Another factor of great importance that has so far received very little consideration in relation to continental water-transfer schemes is climate change. Such schemes, many of which were first conceived in the 1950s and 1960s when very little was known about the impacts of climate change, would require much more than just technological know-how and very strong financial capacities.

Table 5.4   Estimation of Cost of Water Produced or Transported by Different Means, 2002

	Production Cost, According to Various Estimates ($US/m³)	Level of Technological Control	Advantages	Shortcoming
Transfer canal (500 km)	0.8 to 3.0	High	Capacity to deliver large volumes	Huge investments and operation costs, Environmental impact to be assessed
Plastic bags	0.55 (Cyprus) to 1.35 (Greek Islands)	Average	Enables isolated islands or coastal communities to be supplied	Technology needs improvement, Small volumes
Water-carrying ships	1.25 to 1.5	High	Simple technology	Small volumes, Relatively high costs
Iceberg transportation	0.5 to 0.85	Very low	Immense resource to be tapped, Acceptable cost for urban markets	Technology to be much improved for a regular supply
Desalination	From sea water: 0.75 for 40 000 m³/day (Abu Dhabi) 0.85 for 40 000 m³/day (Cyprus), 0.55 for 100 000 m³/day (Tampa Bay)	High	Immense resource to be tapped, Acceptable cost for urban markets, Fast-decreasing operating costs	Large initial investment, Environmental impacts of salt residue

(Continued)

Table 5.4 *(Continued)*

	Production Cost, According to Various Estimates ($US/m³)	Level of Technological Control	Advantages	Shortcoming
Water recycling	0.07 to 1.80	Average to high	Increases resource without developing new sources	The required investments and operating costs are all the higher as the water is more polluted. Rarely socially acceptable as drinking water

Source: Lasserre and Descroix (2011).

Long-term studies on the impacts of climate change, such as changing precipitation and evaporation rates and the thawing of permafrost, would all be required and would render the building of large dams in the Canadian Arctic infinitely more complex than might have been anticipated 50 years ago.

Conclusions

Water diversions are already a fact in both Canada and the United States, and further diversions can be expected, particularly with the development of hydroelectric projects and water-transfer agreements in the southwest.

Continental water-diversion schemes first emerged during the 1950s and 1960s in response to fears of upcoming, persistent droughts, as well as due to the realization (especially in the southwestern United States) that the amount of available water might not be sufficient to keep pace with the forecasted economic and demographic growth. But those forecasts ultimately proved to be wrong, and while the US GDP has sharply increased, water consumption has stagnated. Ultimately, a new emphasis on water efficiency in the United States and a reduced demand for water, coupled with the prohibitive costs of these water diversions, have kept such projects from being realized.

Given the present trend in US water usage and management, as well as the strong opposition that major water-diversion projects face in Canada, it is unlikely that any will be implemented in the foreseeable future. This is particularly the case given that their implementation would require an intensive round of inter-jurisdictional negotiations involving not only the two countries' national governments, but also those of the provinces and states, as well as Indigenous peoples who have been systematically ignored in the past (see chapter 5, this volume). Such schemes would also almost certainly prompt constitutional quarrels in both Canada and the United States. Indeed, such challenges have already been seen on a small scale in California and Arizona with the Colorado Aqueduct and the Central Arizona Project.

Given these challenges, diversions have never been conceived as water exports, but rather as ways to mobilize a useful resource within a single political jurisdiction, whether to produce power, water fields, or supply cities. Water transfers between Canada and the United States remain rare to this day and are mostly limited to a series of transboundary inter-local water agreements between boundary communities. And, even in those cases, conflicts sometimes emerge and make us realize that "water without borders," even at the local scale, is often difficult to achieve. A local, tense hydropolitics is sometimes at play (Trottier 1999), and water conflicts are very often local across the world (Julien and Lasserre 2012). For example, Stanstead, Qc, did not hesitate lately to leverage its control over the water resources it is sharing with Derby Line, VT, to gain political advantage in a debate over wind turbine installation planned on the US side (Associated Press 2012). Such plan was later discarded. Thus, if at times we may witness the "rise of the local" (see chapter 3), the local does not necessarily evolve toward greater communication and cooperation, especially in the case of the Canadian-US border (Lasserre, Forest, and Arapi 2012).

Yet this reality must not be seen as precluding the possibility of future water exports for commercial ends. On the contrary, the scale of existing diversions within Canada makes the nation's resistance to large-scale water exports less credible.

Acknowledgments

This chapter has been revised and updated from a previous version published in Karen Bakker, ed., Eau Canada: The Future of Canada's Water (Vancouver: UBC Press, 2007). The author would like to thank Patrick Forest for his support and his contribution to the writing of this chapter. Patrick Forest helped in his capacity of Postdoctoral Fellow in the Department of Geography at McGill University. In addition, the authors wish to thank Robert Stewart for the linguistic revision and the reviewers for their helpful comments. This research was made possible with the financial support of the Social Sciences Humanities and Research Council (SSHRC) and the Fonds québécois de recherche sur la société et la culture (FQRSC).

NOTES

1 The project has also received some support from powerful engineering, transportation, and financial firms, such as SNC-Lavalin, UMA Engineering, Bechtel, Power Corporation, Rousseau, Sauvé & Warren, Louis Desmarais' Canada Steamship Lines, and the nuclear energy sector.

2 See Chapter 6 by Ralph Pentland, "Key Issues in Canada-U.S. Water Governance," for more discussion on Great Lakes.

3 A diversion can be defined as a water transfer from one watershed to another through the use of human technology, or as a water export related to a transfer of water across the international border through the use of human technology for commercial purposes.

4 The IJC confirmed that it had a veto right on diversions of Great Lakes' water that was divided by the boundary in case an agreement could not be reached between Canada and the United States. Article III provides that new uses, obstructions, and diversions of "boundary waters" that affect the natural level or flow in the other country can be made only with the authority of the country in which the use, obstruction, or diversion will take place and with the approval of the International Joint Commission or pursuant to a special agreement between Canada and the United States (italics mine). This requirement does not apply to certain governmental works that do not materially affect the level or flow of "boundary waters" in the other country, and does not interfere with the ordinary use of water for domestic and sanitary purposes. This does not apply to Lake Michigan, which is wholly located in American territory.

5 The city of Waukesha, a busy suburb of Milwaukee that lies outside of the Great Lakes watershed, developed a proposal to tap into Lake Michigan for its water supply in 2003. This project helped trigger a strong debate regarding Great Lakes water diversions. Despite the Great Lakes – St. Lawrence River Basin Water Resources Compact, the city of Waukesha nevertheless applied in 2010 to the Wisconsin Department of Natural Resources for a diversion of Great Lakes water.

REFERENCES

Artiola, Janick, and Jim Dubois. 1995. "Arizona's Agricultural Ecosystems." *Arizona Comprehensive Environmental Risk Project (ACERP) Report.* http://earthvision.asu.edu/acerp/section2/Chp_03ES.html. Accessed 7 June.

Associated Press. 2012. "Canadians Squall about Vermont Wind Energy Project." *The Guardian,* 17 May. www.guardian.co.uk/world/feedarticle/10247194.

Bakenova, Saule. 2008. "Making a Policy Problem of Water Export in Canada: 1960–2002." *Policy Studies Journal: the Journal of the Policy Studies Organization* 36 (2): 279–300. http://dx.doi.org/10.1111/j.1541-0072.2008.00266.x.

Bakker, Karen, ed. 2007. *Eau Canada: The Future of Canada's Water*. Vancouver: UBC Press.

Barlow, Maude. 2007. *Blue Covenant: The Global Water Crisis and the Coming Battle for the Right to Water*. Toronto: McClelland & Stewart.

Bingham, J. and Western States Water Council, 1969. A review of inter-regional and international water transfer proposals. Salt Lake City, Western States Water Council.

Bocking, Richard. 1972. *Canada's Water: For Sale?* Toronto: James Lewis & Samuel.

Bourassa, Robert. 1985. *Power from the North*. Toronto: Prentice-Hall.

Boyer, Marcel. 2008. "L'exportation d'eau douce pour le développement de l'or bleu québécois." *Les Cahiers de recherche de l'Institut économique de Montréal (IEM)*: 1–32. http://www.iedm.org/files/cahier0808_fr.pdf.

Cline, Harry. 2003. "Flexible Management for Different Seasons." *Western Farm Press*, 27 March. http://westernfarmpress.com/flexible-management-different-seasons.

Consulate General of Canada in Chicago. 1982. "The Great Lakes Water Diversion Issue." Ottawa: Ministry of External Affairs, 23 April, SFAX 097/Restricted.

Council of Great Lakes Governors (CGLC). 2005. "The Great Lakes-St. Lawrence River Basin Water Resources Compact Water Without Borders and The Great Lakes-St. Lawrence River Basin Sustainable Water Resources Agreement." www.cglg.org/projects/water/CompactImplementation.asp.

Day, John C., and Frank Quinn. 1992. "Water Diversion and Export: Learning from the Canadian Experience." Department of Geography Publication 36, Canadian Association of Geographers, University of Waterloo.

The Economist. 2003. "Pipe Dreams." 11 January.

FLOW, 2010. "Public Opinion – IPSOS Reid survey", www.flowcanada.org/security/public-opinion ·

Forest, Benjamin, and Patrick Forest. 2012. "Engineering the North American Waterscape: the High Modernist Mapping of Continental Water Transfer Projects." *Political Geography* 31 (3): 167–83. http://dx.doi.org/10.1016/j.polgeo.2011.11.005.

Forest, Patrick. 2010a. "A Century of Sharing Water Supplies Between Canadian and American Borderland Communities." Program on Water Issues, Munk School of Global Affairs, University of Toronto. http://munkschool.utoronto.ca/assets/files/Publications/forrestpaper_2010.pdf.

Forest, Patrick. 2010b. "Inter-Local Water Agreements: Law, Geography, and NAFTA." *Cahiers de droit* 51 (3–4):749–70.

Francoeur, Louis-Gilles. 2005. "Conférence des maires des Grands Lacs et du Saint-Laurent – Les eaux des Grands Lacs seront protégées contre les États assoiffés. Les dérivations massives seraient interdites, sauf exceptions." *Le Devoir*, 28 May, A5.

Francoeur, Louis-Gilles. 2006. "La Politique Nationale de l'eau du Québec : une œuvre inachevée." In *Les Politiques de L'eau. Grands Principes et Réalités Locales*, ed. Alexandre Brun and Frédéric Lasserre, 47–68. Québec: Presses de l'Université du Québec.

Gingras, Pierre. 2009. "The Northern Waters: A Realistic, Sustainable and Profitable Plan to Exploit Quebec's Blue Gold." Montréal Economic Institute press release. www.iedm.org/files/juillet09_en.pdf.

Gingras, Pierre. 2010. *L'eau du Nord. Un projet réaliste, durable et rentable pour exploiter l'or bleu du Québec*. Montréal: Marcel Broquet.

Gleick, Peter H. 2001. "A Look at Twenty-first Century Water Resources Development." *Water International* 25 (1): 127–38. http://dx.doi.org/10.1080/02508060008686804.

Gleick, Peter. 2010. "Water Use in the United States has Leveled Off: New Remarkable Numbers Released." http://blog.sfgate.com/gleick/2009/10/29/water-use-in-the-united-states-has-leveled-off-new-remarkable-numbers-released/. Accessed 30 May 2011.

Government of the United States of America. 2000. *Water Resources Development Act (WRDA)*. Amended 11 December 2000, P.L. 99–662, Title XI, 1109, 100 Stat. 4230.

Hanak, E., J. Lund, A. Dinar, B. Gray, R. Howitt, J. Mount, P. Moyle, and B. Thompson. 2009. "California Water Myths." Public Policy Institute of California. www.ppic.org/content/pubs/report/R_1209EHR.pdf.

Holm, Wendy, ed. 1988. *Water and Free Trade. The Mulroney Government's Agenda for Canada's Most Precious Resource*. Toronto: James Lorimer & Company.

Humlum, Johannes. 1969. *Water Development and Water Planning in the Southwestern United States*. Aarhus: Aarhus University Press.

Hydro-Québec. 2001. *Synthèse des connaissances environnementales acquises en milieu nordique de 1970 à 2000*. Montréal: Hydro-Québec.

International Joint Commission (IJC). 2000. "Protection of the Waters of the Great Lakes – Final Report to the Governments of Canada and the United States." www.ijc.org.

Johansen, David. 2010. "Bulk Water Removals: Canadian Legislation." Publication 02-13-E, Ottawa: Parliament of Canada. www.parl.gc.ca/content/LOP/ResearchPublications/prb0213-e.pdf.

Julien, Frédéric, and Frédéric Lasserre. 2012. "Anthropomorphisme et hydrocentrisme dans la thèse des guerres de l'eau: les racines d'un improbable scénario du pire." *Bulletin de l'Association des Géographes Français (BAGF) – Géographies* 1: 54–73.

Katz, Diane. 2010. "Making Waves: Examining the Case for Sustainable Water Exports from Canada." Vancouver: Fraser Institute. www.fraserinstitute.org/uploadedFiles/fraser-ca/Content/research-news/research/publications/Case-for-sustainable-water-export.pdf.

Keating, Michael. 1986. *To the Last Drop. Canada and the World's Water Crisis*. Toronto: Macmillan.

Klymchuk, D. 2001. Water Exports-A Manitoba Bonanza? Backgrounder 7, Frontier Centre for Public Policy, Winnipeg.

Lasserre, Frédéric. 1998. *Le Canada d'un mythe à l'autre. Territoire et images du territoire*. Lyon, Montreal: HMH/Presses Universitaires de Lyon.

Lasserre, Frédéric. 2001. "L'Amérique a Soif. Les Besoins en Eau de l'Ouest des États-Unis Conduiront-ils Ottawa à Céder l'Eau du Canada?" *Revue Internationale d'Études Canadiennes/International Journal of Canadian Studies. Revue d'Etudes Canadiennes* 24:196–214.

Lasserre, Frédéric. 2002. "L'Eau, la Forêt, les Barrages du Nord du Québec. Un Territoire Instrumentalisé?" In *Le Territoire Pensé: Géographie des Représentations Territoriales*, ed. F. Lasserre and A. Lechaume. Sainte-Foy: Presses de l'Université du Québec.

Lasserre, Frédéric. 2005a. "Introduction." In *Les Transferts d'Eau Massifs. Outils de Développement ou Instruments de Pouvoir?* ed. F. Lasserre, 1–35. Sainte-Foy: Presses de l'Université du Québec.

Lasserre, Frédéric. 2005b. "La Continentalisation des Ressources en Amérique du Nord. L'ALENA oblige-t-elle le Canada à céder son eau aux États-Unis? " In *Les Transferts d'Eau Massifs. Outils de Développement ou Instruments de Pouvoir?* ed. F. Lasserre, 463–88. Sainte-Foy: Presses de l'Université du Québec.

Lasserre, Frédéric. 2005c. "Les Projets de Transferts Massifs Continentaux en Amérique du Nord. La Fin de l'Ère des Dinosaures?" In *Les Transferts d'Eau Massifs. Outils de Développement ou Instruments de Pouvoir?* ed. F. Lasserre, 489–532. Sainte-Foy: Presses de l'Université du Québec.

Lasserre, Frédéric. 2005d. "Great Lakes Governors' Agreements." In *Water Encyclopedia. Water Quality and Resource Development*, tome 3 *Oceanography, Meteorology, Physics and Chemistry, Water Law, and Water History, Art, and Culture*, ed. Jay H. Lehr and Jack Keeley, 544–47. Hoboken, NJ: Wiley.

Lasserre, Frédéric. 2006. "Drawers of Water: Water Diversions in Canada and Beyond." In *Eau Canada. The Future of Canada's Water*, ed. Karen Bakker, 143–62. Vancouver: UBC Press.Lasserre, Frédéric, Patrick Forest, and Enkeleda Arapi. 2012."Politique de sécurité et villages-frontière entre États-Unis et Québec," *Cybergéo: European Journal of Geography* 595. http://cybergeo.revues.org/25209. Accessed 3 March. doi: 10.4000/cybergeo.25209.

Las Vegas Sun. 2005. "Desalination May Be Solution to Water Woes of LV Valley." 23 October.

Las Vegas Sun. 2006. "Desalination Still Years Away for West." 29 June.

Las Vegas Sun. 2008. "A Desalination Plant Is Not the Immediate Answer to Southern Nevada's Water Needs." 25 March.

Lee, Philip. 2001. "The Wellspring of Life, or Just a Commodity. Cost, Not Emotion, Likely to Kill Export Idea." *Ottawa Citizen*, 16 August.

Maich, Steve. 2005. "America is Thirsty." *Maclean's Magazine*, 28 November.

Matheson, Scott. 1984. "The Sharing of Water on a Continental Basis." *Futures in Water, Proceedings of the Ontario Water Resources Conference*, Toronto, 12–14 June.

Micklin, Philip. 1985. "Inter-Basin Water Transfers in the United States." In *Large-Scale Water Transfers: Emerging Environmental and Social Experiences*, ed. Genady Golubev and Asit Biswas, 37–66. Oxford: Tycooly.

Moss, Frank E. 1967. *The Water Crisis*. New York: Praeger.

Nanos, Nik. 2009. "Canadians Overwhelmingly Choose Water as Our Most Important Natural Resource." *Policy Options* 30 (7): 12–15.

Nicholson, N. 1979. *The Boundaries of the Canadian Confederation*. Carleton Library 115, Ottawa.

Nilsson, C., C.A. Reidy, M. Dynesius, and C. Revenga. 15 Apr, 2005. "Fragmentation and Flow Regulation of the World's Large River Systems." *Science* 308 (5720): 405–8. http://dx.doi.org/10.1126/science.1107887. Medline:15831757

Pentland, Ralph, and A. Hurley. 2010. "Canadian Water Issues Council (CWIC) Response to Bill C-26–Letter to Foreign Affairs Minister Lawrence Cannon." Program on Water Issues, Munk School of Global Affairs, Toronto, 2 June.

Postel, Sandra, and Brian Richter. 2003. *Rivers for Life: Managing Water for People and Nature*. Washington, D.C.: Island Press.

Quinn, Frank, J.C. Day, M. Healey, R. Kellow, D. Rosenberg, and J.O. Saunders. 2004. "Water Allocation, Diversion and Export." In *Threats to water availability in Canada*, Environment Canada, NWRI Scientific Assessment Report Series No. 3 and ACSD Science Assessment Series No. 1. Burlington: National Water Research Institute.

Quinn, Frank. 2007. "Water Diversion, Export, and Canada-U.S. Relations: a Brief History." Munk Centre for International Studies Briefings, University of Toronto. http://powi.ca/wp-content/uploads/2012/12/Water-Diversion-Export-and-Canada-US-Relations-A-Brief-History-2007.pdf.

Rowe, Craig, 2011. "When it Comes to Importing Water, Nothing Seems Too Extreme." *High Country News*, 29 September.

San Diego County Water Authority. 2002. "Diversification." *Annual Report*, 10–11.

Schroeter, H., A.D. Arthur, and W.S. Baskerville. 2005. "Establishing Environmental Flow Requirements for Big Creek." Toronto: Conservation Ontario. http://www.conservation-ontario.on.ca/projects/pdf/Long%20Point.pdf.

Simon, Paul. 1998. *Tapped Out: The Coming World Crisis in Water and What We Can Do About It*. New York: Welcome Rain Publishers.

Solomon, Lawrence. 1999. "Water Export Threat is Trickling Away." *National Post*, 30 November.

Southern Nevada Water Authority (SNWA). 2005. *2006 Water Resource Plan*. Las Vegas, 26–27.

Special Subcommittee on Western Water Development of the Committee on Public Works. 1964. *Western Water Development: A Summary of Water Resources Projects,*

Plans, and Studies Relating to the Western and Midwestern United States. Washington, D.C.: US Government Printing Office.

Sprague, John B. 2007. "Great Wet North?" In *Eau Canada: the Future of Canada's Water*, ed. Karen Bakker, 23–35. Vancouver: UBC Press.

Sproule-Jones, M., C. Johns, and B.T. Heinmiller, eds. 2008. *Canadian Water Politics: Conflicts and Institutions*. Montreal: McGill-Queen's University Press.Thompson, Jerry. 1999. "American Thirst, Canadian Water." Raincoast Storylines. CBC, Discovery Channel.

Trottier, Julie. 1999. *Hydropolitics in the West Bank and Gaza Strip*. Jerusalem: PASSIA.

United States Geological Survey (USGS). 1998, 2004, 2009. *Water Use in the United States*. Washington, D.C.

US Water News Online. 2004. "Bush, Kerry Agree on Water Diversions – but Fight about it Anyway." November. www.uswaternews.com/archives/arcpolicy/4bushkerr11.html. Accessed 7 June 2011.

Worster, Donald. 1985. *Rivers of Empire. Water, Aridity, and the Growth of the American West*. New York: Oxford University Press.

Wright, Jim. 1966. *The Coming Water Famine*. New York: Coward-McCann.

Wright, Steven, and Jonathan Bulkley. 2002. "Diversion and Consumptive Uses of Great Lakes Waters: A Framework for Decision-Making." Proceedings from the symposium titled, *Our Challenging Future: A Review of the State of Great Lakes Research*, University of Michigan, 5–6 November.

6 Key Challenges in Canada-US. Water Governance

RALPH PENTLAND

In the new global economy, the industrialized world appears to be operating under the assumption that global economic growth, the promotion of democratic systems, and the encouragement of international trade and investment will produce a "virtuous cycle" of wealth generation, social advance, and eventually, ecological protection (Pentland 2010b). In other words, there is an implicit assumption that society can delay dealing with the most egregious forms of ecological decline until the consumptive desires of the wealthy, as well as the consumptive needs of the poor, are met.

Whether by accident or design, that philosophical direction has had some very fundamental implications for governance and the Canada-US water relationship. We can no longer assume that this relationship is limited to the waters these two countries share. For example, Canada has become one of the last "reliable" suppliers of US petroleum products. This has created huge incentives for rapid, and some would say reckless, expansion of oil sands development, despite its large-scale impacts on the Athabasca River and its potentially significant impacts downstream on the Mackenzie River.

Aside from the waters we share and those affected by resource extraction, broader US environmental policies are now effectively deciding the fate of many other Canadian waters. For example, US policies on climate change and chemicals management have profoundly affected trends in both the quality and quantity of waters that support the lives and livelihoods of all Canadians. In the not-too-distant future, influences from well beyond North America, such as food shortages and widespread environmental stresses caused by climate change, could result in both significant challenges and opportunities for North Americans.

The bottom line is that the "virtuous cycle" is at least temporarily morphing into a "vicious cycle" of increasing global-scale competition for scarce resources, ecological decline, environment-related human-health issues, and economic shocks. Canada and the United States, along with the rest of the world, will have to become ingenious at coping with and adapting to an already embryonic "vicious cycle," until conventional wisdom advances enough to make room for more fundamental solutions. In the Canada-US water and environmental context, the two countries need to move beyond merely coping by proactively promoting the development and implementation of those more fundamental solutions.

In the following sections, the potential for moving toward more lasting solutions is examined: first, with respect to a few key water issues such as water export, climate change, threats to groundwater, and the changing face of water pollution; second, with respect to institutional concerns; and finally, with respect to the possibility of turning the challenges into opportunities.

Unresolved and Emerging Water Issues

Until about 1965, there was a broad societal trend in both the United States and Canada concerning nation-building. In the context of Canada-US water relations, the highest-profile manifestation of that consensus was through major projects of mutual advantage, such as the St. Lawrence Seaway and Columbia River hydropower development (for more discussion of the Columbia River, see chapter 7). Between 1965 and about 1990, it was suddenly realized that one person's effluent was another person's intake, and that the public interest did not always coincide with the pursuit of private interest. This awareness led to many cross-border impact situations, such as the Garrison Diversion proposal in North Dakota. It also led to more systemic technical approaches, including a more systemic approach to Great Lakes water-level regulation and an ecosystem approach under the Great Lakes Water Quality Agreement.

Since about 1990, globalism and competitiveness have clearly dominated the public-policy agenda in both Canada and the United States, prompting governments to relax regulatory regimes and de-emphasize environmental priorities. As a result, environmental and water management capacity at all levels of government have been significantly hollowed out. Under those circumstances, one would expect Canada-US water relations to suffer, and they certainly have. This has been most apparent in the water quality sector, where earlier gains in the Great Lakes region have been largely reversed, and new cross-border impacts, such as those in Lake Winnipeg, are becoming very serious. Looking ahead, there will continue to be numerous skirmishes of a local nature, but the more serious bilateral water issues are likely to be rooted in broader national and international issues and policies. These two countries will have to become fairly ingenious in dealing with a water relationship that extends well beyond just the watersheds they share. Some of the most significant issues for the future will likely include water exports, climate change, emerging threats to groundwater, and the changing face of water pollution.

The differing roles of the senior governments in Canada and the United States are a major institutional consideration for the governance of shared waters. The Canadian constitution assigns authority over most natural resources to the provinces, while the US federal government plays a more dominant role than state powers do in terms of water management. This system has some very practical implications. For example, with respect to pollution, the very timid approach of Canada's federal government, combined with nonbinding, consensus-based decision-making by the Canadian Council of Ministers of the Environment, virtually ensures a poor Canadian performance relative to that of other industrialized nations. By contrast, the US federal government sets national policy and binding, countrywide floor standards, which the states often exceed. Therefore, it should not be surprising that in virtually all comparisons of emissions data, Canada performs much poorer than its southern neighbours.

Canada is also a work in progress with respect to foreign relations in general. Canada only took control of its foreign affairs policy in 1931 and repatriated its constitution from Britain in 1982. While treaties approved by the US Senate are the law of the land for all levels of US government, similar treaties in Canada must be implemented by federal or provincial legislation, depending on their respective authorities. An interesting example is the Columbia River Treaty (see chapter 7, this volume). This treaty was agreed upon by the two countries in 1961, but did not take effect until three years later, after the premier of British Columbia successfully defied Ottawa regarding the terms of selling its hydropower (Quinn 2006). Such interactions have resulted in a trend in which the Canadian federal government tends to wait for provinces to act, and then reacts, often belatedly.

In recent years, this kind of uncertainty has been overcome to some extent by way of direct state-provincial negotiations, with the two federal governments sanctioning the agreements after the fact as necessary. An earlier example was cost-sharing for the Rafferty and Alameda dams in Saskatchewan, which was negotiated by Saskatchewan and North Dakota, and subsequently supported by a Canada-US agreement regarding water sharing in the Souris River Basin. A more recent example is the agreement between eight US states and two Canadian provinces regarding diversions and consumptive uses in the Great Lakes – St. Lawrence region. That agreement was subsequently passed into law in the United States in the form of an interstate compact. This approach has many advantages, but it also has its inherent risks, namely that provinces and states may not always understand their rights under the Boundary Waters Treaty, and they may not always act in the broader national interest.

Water Exports

Even as early as the 1970s, public concern about the bilateral water relationship started to expand to waters beyond the boundary region and into the "back country" (for a more detailed discussion, see chapter 5 of this volume). For example, there was growing interest in the export of electricity and emerging issues such as acid rain. Of particular interest, however, was the spectre of water shortages that occurred in the southwestern states in 1963. At this time, the US Supreme Court resolved a legal battle between Arizona and California over their respective shares in the Colorado River, which precipitated a political struggle to capture new water supply outside the Colorado River Basin states. This in turn led to increased demand for federal legislation to support importation of water from other western regions, in addition to a proliferation of schemes from the private sector to make the Columbia, the Missouri, and the Great Lakes tributary to the Colorado River (Quinn 2007).

None of the international schemes volunteered by the private sector at that time or since has ever been supported, or even considered seriously, by either federal government. Nevertheless, the issue of water export, more than four decades later, remains very much alive. As Lasserre discusses in chapter 5, Canadians continue to overwhelmingly fear and reject the notion of trading water as they do any other natural resource, no matter which side of the boundary originates the proposals. Canadian engineers have increasingly added their own water-transfer schemes to the list, which was at first dominated by their US counterparts.

During the 1960s, there were at least nine such schemes on the table. By far the largest was the North American Water and Power Alliance (see box 6.1), proposed by a California engineering firm. It involved diverting flows from the Mackenzie and Yukon River Basins southward through the Rocky Mountain Trench and into the United States. This diversion's stated purpose was to provide irrigation water to the southwestern United States, generate hydroelectricity, and create navigation routes. It featured a huge new network of waterways throughout the western half of the continent.

Box 6.1   The North American Water and Power Alliance (NAWAPA)

Ralph Pentland

In 1964, the Ralph M. Parsons Company first proposed this very ambitious continental-water scheme to a special subcommittee of the US Senate chaired by Senator Frank Moss from Utah. Its proponents claimed at the time that it would ensure adequate water supplies to meet the needs of the entire continent for the next 100 years. The central feature of the scheme would be a man-modified reservoir 500 miles long, 10 miles wide, and 300 feet deep in the southern end of the natural gorge known as the Rocky Mountain Trench in British Columbia. The reservoir would be fed from a series of dams that would trap water from several rivers running through wilderness areas in Alaska and the Yukon Territory.

To the east, a 30-foot canal would carry water to Lake Superior, providing irrigation water to the Canadian and US Great Plains Regions, and ultimately, creating a navigation canal from the Great Lakes to Howe Sound, British Columbia. South of the Trench, water would be lifted to the Sawtooth Reservoir in Montana, before flowing southward to the western and southwestern US states and Mexico. Overall, the original plan called for no fewer than 369 separate projects.

Most water experts in both Canada and the United States dismiss proposals of this kind as being impractical. South of the border, there appears to be less interest in importing water than at any time in the recent past. Southwestern states have been rebuffed in turn by their better-watered neighbours in the Pacific Northwest, the lower Mississippi, and the Missouri and Great Lakes Basin states. This does not seem to have caused a problem, however, as much as a change of direction. Water supplies within the Southwest are not running out; they are of necessity being used more efficiently.

According to the latest data on water use, Americans continue to withdraw less water from surface and underground sources than they did in 1975, despite significant population and economic growth; they have, in effect, broken the link between population and water use. Nationally, per capita use has dropped nearly 30 per cent since 1975, as Americans produce more goods and services more efficiently.

The second largest plan was the massive GRAND Canal scheme, first proposed in 1959; it has repeatedly resurfaced in a wide variety of forms ever since. The version that caused the most controversy involved the construction of a dike across James Bay, which would trap the rivers flowing into it from Ontario and Quebec and send 17 per cent of their flows southward. This flow, roughly equivalent to 30 per cent of the Great Lake's discharge, would be channelled through river valleys, reservoirs, and pumping stations for about 640 kilometres and raised 300 metres over the Canadian Shield before being discharged into the Ottawa River. From there, water would be diverted to Lake Huron through Lake Nipissing and the French River, travelling another 180 kilometres. Water from the Great Lakes would then be diverted to the southwestern region of the United States and the drier regions of western Canada (Pearse et al. 1985).

Massive schemes of this type were never much more than lines on a map, and water experts never took them seriously. However, they have at times been taken seriously by some politicians and have raised alarm bells with Canadian citizens. To soothe these alarms while still keeping water-export prospects alive, several Canadians have since suggested more modest, but equally impractical ideas, including pipeline proposals from northern Manitoba and northern Quebec, and scaled-back versions of the GRAND Canal scheme. There have also been a variety of marine tanker proposals, which tend to be somewhat more practical and palatable than the overland schemes, but which have not gained any traction to date.

Why has bulk-water export not taken off despite a half-century of intense lobbying on its behalf, and at least some cautious, behind-the-scenes big business and political support? Basically, the answer can be attributed to a coincidence of four interrelated factors:

1. Each new proposal has always been greeted by hostility from an overwhelming majority of Canadians, which has limited the political options.
2. The larger-scale proposals were never taken seriously by water experts in either country because they were clearly impractical for a variety of socioeconomic, environmental, political, and legal reasons.
3. There have always been a number of misconceptions about potential markets.
4. In the Canadian context, long-distance, large-scale bulk-water export schemes would quite simply represent bad water policy because they ignore the ecological integrity and economic potential of the donor regions.

Climate Change

The changes are clear: greenhouse gas emissions are now more than 40 per cent above 1990 levels, glaciers are visibly receding, ice on North American lakes and rivers is disappearing, impacts in the Arctic are already becoming obvious, and extreme weather events are becoming more common (Bruce 2010). Climate change will have implications for both the quantity and quality of boundary and transboundary waters. Some water-quality impacts have already been observed. In lakes and reservoirs, higher water temperatures are leading to longer periods of summer stratification, as well as reductions in dissolved oxygen levels in these water bodies and in rivers. Low levels of oxygen cause stress on aquatic animals, including coldwater fish and the insects and crustaceans on which they feed. Pollution from land use is being amplified by increases in precipitation intensity and longer periods of low flow. This increased pollution, coupled with higher temperatures, is resulting in blooms of harmful algae and bacteria (Pentland 2010a).

Emerging Groundwater Issues

Serious transboundary groundwater issues have not arisen to date, but that may very well change in the coming decades. Some of the activities likely to trigger this eventuality are recent technological developments related to carbon capture and storage, as well as hydraulic fracturing, or "fracking," in shale regions for natural gas.

In terms of the first major concern, large-scale deployment of carbon capture and storage could result in leaks over time into the air or groundwater. These leaks, which would pose a threat to both human and ecosystem health, could also pose a variety of largely unquantified risks to groundwater systems located above carbon-storage basins throughout western North America (Forum for Leadership on Water 2010).

As for the second concern, shale deposits are attracting the attention of governments and energy industries all around the world because technological advances have made it economically viable to extract the natural gas locked in deep shale formations. Conceptually, fracking is simple, but its execution is an engineering feat. It amounts to using brute force to crack open pathways in dense rock, making it easier to extract shale gases. A combination of horizontal drilling and fracking technologies has caused global natural gas production to soar. However, the risks associated with this activity are still poorly understood. In certain formations, the shale is characterized by vertical cracks, and because shale gases are typically overpressurized and the fracking process further increases the pressure, this process can open pathways upward to freshwater. Three major concerns are related to the potential impacts of fracking operations on water quality: the specific chemicals used, groundwater contamination, and contamination from the large volumes of wastewater produced in these operations (Pentland 2010c).

Given the paucity of groundwater information in transboundary regions and a lack of standards in both countries, but especially in Canada, the push to accelerate both carbon capture and storage and hydraulic fracturing for shale-gas exploitation (fracking) could pose very real risks to shared groundwater systems. Moving ahead with either or both of these methodologies would be highly risky without strong regulation, clearly established liabilities, effective oversight, sound science, and transparent decision-making processes.

Regardless of the risks, our shared energy thirst will almost surely cause us to move ahead with at least shale-gas fracking on a very large scale – and without the necessary safeguards in place.

The Changing Face of Water Pollution

The most alarming aspect of the changing face of pollution is a growing awareness that newer forms of air and water contamination may be significantly affecting human health. Before the turn of the century, much of the attention focused on bioaccumulative toxic substances (PBTs) and persistent organic pollutants (POPs), which can move through food chains and thereby threaten top predators in an ecosystem, including fish-eating birds and mammals. Some of these PBTs and POPs include industrial chemicals in the polychlorinated biphenyl (PCB) family, pesticides, dioxins and furans, plasticizers, flame retardants, bisphenol-A, and polyaromatic hydrocarbons. Many PBTs and POPs are distributed through atmospheric transport, and even those that have been strictly controlled or banned in Canada continue to be detected in fish and wildlife (Hewitt and Servos 2001).

The changing face of water pollution has taken another unexpected and very troubling turn over the past decade. The presence of pharmaceuticals and personal-care products (PPCPs) in the environment gained international attention after a 2000 report by the US Geological Survey provided a comprehensive water-quality assessment of 139 streams throughout 30 states. In some streams, especially those where the source of water consisted primarily of treated effluent, concentrations of PPCPs were unusually high. This report raised a number of questions about human-health risks, potential contamination of groundwater, analytical validity, and compound identification and classification (Kolpin et al. 2002).

The detection of pharmaceuticals and related products in water systems quite appropriately generates considerable concern, simply because many of these products are specifically designed to have a biological effect on humans, animals, and plants. Pharmaceuticals often contain chemical compounds that can affect the endocrine system by altering, mimicking, or impeding the function of hormones. Over the past decade, scientists and regulators have become increasingly alarmed about the subtle effects that exposure to even very low levels of PPCPs may have on aquatic organisms, and subsequently, as they move up the food chain, on human health.

Pharmaceuticals include prescription drugs, veterinary drugs, and over-the-counter medicines. Personal-care products cover a broad spectrum including cosmetics, hair products, sunscreens, fragrances, antibacterial soaps, and vitamins. PPCPs enter water bodies through human excretions and the disposal of unused prescriptions in municipal wastewater, via animal waste and fertilizers in agricultural runoff, and through aquaculture operations. Some PPCPs are not removed by traditional sewage treatment, and traces have been found in drinking-water supplies.

Banning or strictly controlling known endocrine disrupting substances is clearly important, and some progress is being made in that regard. However, these efforts only deal with a relatively small portion of the problem. Our less-than-comprehensive chemical-management systems in both Canada and the United States continue to be a source of considerable concern. Even if our chemical-management systems were comprehensive, complex mixtures of substances produced in multiple industrial processes, with causative agents that are difficult to isolate and identify, would continue to confound our best efforts. Both countries will have to upgrade their approaches to chemicals management, and more generally, to toxic-emissions management in the coming decades, and will have to do so in concert to avoid cross-border tensions.

Emerging Approaches to Emerging Issues

Discussions of international security are increasingly focusing on water and environmental issues. Is there going to be enough food and clean water in the future to meet society's needs? How will we plan a collective response to catastrophic extremes of weather and climate? For Canada and the United States, the answers to these questions are profoundly affected by each other's policies because our futures are bound together by geography and a highly integrated economy. With that overriding notion in mind, this section reflects on some possible approaches to dealing with each of the main issues raised above.

Water Export

Given current trends in US water use and management, there is no reason to expect demands for bulk-water export in the foreseeable future (for a detailed discussion, see chapter 5, this volume). However, there is one wild card, namely climate change. The US population continues to migrate southward and westward, particularly to coastal regions and other parts of Texas and California. This shift in population puts citizens on a collision course with the storms, rising sea levels, and extended droughts that are associated with climate warming. Should the United States find itself in desperate need for water in the future, and should Canada refuse to enter into an export agreement, what would prevent the United States from simply taking a disproportionate share of waters along the international boundary?

There are at least three obvious possible locations for water transfer. The first is the Great Lakes, and more specifically, the Chicago Diversion, which is specifically exempt from the state-provincial agreement prohibiting removals of water from the Great Lakes Basin. A second example is in the Red River Basin, where a draft 2005 US Bureau of Reclamation study included the option of diverting water from the shared Lake of the Woods to the US portion of the basin. A third example is in the St. Mary – Milk region of the Great Plains (see also chapter 8, this volume), where US interests have been demanding a reexamination of the long-standing international water-apportionment agreement.

The bulk-water export question is largely in hand in Canada. The federal and provincial governments agree that water should generally be kept within its natural watersheds, with minor and well-defined exceptions. This fundamental premise is also central to state-provincial agreements in the Great Lakes – St. Lawrence region and to existing legal regimes in most Canadian provinces. Unfortunately, this principle has not yet been universally accepted in the United States and continues to raise unnecessary conflicts, especially in the Midwest. This is one area where policy leadership from Canada could conceivably result in a coordinated approach with benefits to both countries.

As mentioned earlier, even though bulk-water export is only a remote threat in the foreseeable future, competition for resources in shared basins will inevitably continue. Canada can minimize that competition in a number of ways. First, it needs to build enough front-line capacity to analyse and bring transparency to all proposals to take water. Second, it needs institutions and individuals who can bargain effectively with an aggressive, but law-abiding neighbour. Third, it needs to work with the US public and political leaders to demonstrate that their own long-term interest would be best served by promoting water-use efficiency and other local solutions. Fourth and finally, it needs domestic resolve to manage its own waters in an exemplary fashion to lead by example (Pentland and Hurley 2007).

Climate Change

Climate change will raise several significant challenges. Regarding water apportionment, most water-sharing arrangements between the two countries are now based on a percentage of preproject flow. As long as the countries continue to follow that approach and the percentages remain the same after climate change, that challenge should be manageable. The more difficult challenges will relate to management in a state of hydrologic uncertainty and nonstationarity.

Continual adaption will obviously become the name of the game, but we also should be following certain "no regrets" policies, since they will be beneficial whether or not our longer-term hydrologic prognostications are correct. For example, we should:

- Seriously question the real viability and sustainability of any proposed new irrigation systems, and make the existing ones more water efficient.
- Develop more flexible reservoir operation and other water management methodologies.
- Keep water within its natural watersheds to maximize resilience to deal with unforeseen circumstances.
- Discourage new development on flood plains and along shorelines that are susceptible to erosion.
- Price water services in a way that encourages water conservation.
- Do less subsidizing and more taxing of environmentally damaging activities, and practice user and polluter pay principles to the extent practicable.
- Engage and enable local entities, and encourage responsible local decision-making within constraints that protect the broader public interest.
- Reverse the serious degradation of our environmental policy, science, and planning capacities, which we have witnessed in recent years.

There are also things we should not do. For example, we should not try to counter uncertainty with heroic engineering measures, like some of the continental plumbing schemes proposed mostly by Canadian dreamers. The keys to success in a very interconnected and uncertain future will be flexibility and resilience. The danger of overreacting or reacting in inappropriate ways with massive structural measures is significantly greater than the danger of underreacting.

Emerging Groundwater Issues

Currently, it is unclear whether carbon capture and storage (CCS) will contribute to the ultimate climate-change solution or if it merely represents political folly. What is clear is that Canada and the United States do not have sufficient scientific evidence to predict accurately the consequences of large-scale CCS. Canada is particularly vulnerable because of its much weaker groundwater-monitoring programs. In 2008, the Council of Canadian Academies decried Canada's ignorance of its groundwater resources despite 10 million Canadians relying on drinking water that comes from beneath their feet. According to the council, combining an aggressive CCS regime with our lack of knowledge about groundwater could lead to disaster (Council of Canadian Academies 2009).

What is equally clear is that regulatory regimes are still woefully weak in both countries, although much more progress is being made south of the border where the United States is taking much more decisive and transparent action than Canada. Since CCS is vastly different from other underground storage activities, the US EPA has introduced proposals for a new class of injection wells specific to CCS, covering site selection, monitoring, well construction, decommissioning, and site closure. There is nothing even remotely equivalent at the federal level in Canada (Thompson 2009).

Differences in the two countries' science and regulatory regimes raise serious CCS concerns, both domestically and binationally. Domestically, potentially enormous risks to human health and ecosystems exist. Binationally, if injections in any Canadian province were to lead to transboundary harm, the potential for significant international liability is clear. That liability would attach, as a matter of international law, to the government of Canada, with the costs ultimately borne by all Canadian taxpayers. CCS may become stalled based simply on a lack of economic viability. However, if it does proceed on even a moderate scale, an urgent need exists to upgrade the relevant science and regulatory regimes in both countries, but especially in Canada.

In some ways, shale-gas fracking is a much more imminent and potentially serious issue because it is already proceeding on a relatively large scale in the United States and is at the same time gaining momentum in Canada. As with CCS, the United States is moving forward much more aggressively than Canada in improving the state of knowledge and regulatory regimes. In March 2010, the US EPA announced that it would conduct a "comprehensive research study" on the potential adverse impacts of hydraulic fracturing on water quality and human health. Several US state and federal government initiatives also point to increasing regulation of the shale-gas industry, including the possibility of increased "no-go" or at least "harder-to-go-into" zones, where gas companies will have to meet higher standards before drilling occurs (Parfitt 2010). Once again, there isn't anything even closely equivalent happening in Canada, but there is an urgent need to take action for both domestic and cross-border reasons.

The Changing Face of Water Pollution

With regard to water pollution, especially toxic releases, several international comparisons suggest that both Canada and the United States perform poorly relative to most other industrialized nations. Even more disconcerting is the fact that head-to-head Canada-US comparisons show that Canada's performance is significantly worse than that of the United States (Weibust 2009). While international comparisons of this type are far from flattering to Canada, they do nevertheless offer a few glimmers of hope. First, even though the United States is not a stellar performer, it does outperform Canada by a significant margin. Since Canada is their closest neighbour and largest trading partner, shares so many common river basins with them, and tends to mimic many of their policies over time, one would logically expect that the United States will eventually drag Canadian performance upward. That might be expected to begin, for example, in the Great Lakes region under a renewed water-quality agreement.

Second, we know from many earlier experiences that whenever a very strong international consensus emerges that some water-pollution issue has become especially urgent, the international community, including Canada and the United States, has always responded appropriately. For example, both countries have effectively reduced lead emissions, the use of contaminants such as DDT and PCBs, releases of dioxins and furans from pulp and paper mills, and chemical discharges from petroleum refineries.

The most comprehensive chemical-management system in the world is most likely REACH, the European Union's chemical registration, evaluation, and authorization system. Under EU rules, any company that manufactures or imports more than one metric ton of a substance is required to register the chemical in a central database. Registration includes details about the substance's properties, its uses, and safe ways to handle it. A new European chemicals agency is responsible for reviewing registrations and making nonconfidential information available to the public. Each registered chemical's dossier is examined to evaluate its compliance with registration requirements and the existing test data related to the substance. If a human health or environmental risk is even suspected, further testing can be required. Official authorization is required to continue the use of substances determined to pose serious and irreversible risks.

Note that even before the introduction of REACH in 2007, Europeans were far ahead of North Americans in terms of chemicals management. For example, unlike Canada and the United States, Europe prohibits the use of arsenic, antibiotics, and hormones in livestock to foster accelerated growth and has much stronger standards governing asbestos, fire retardants, pesticides, cosmetics, and personal care products. Requirements in some individual European countries are even more stringent. For example, Sweden plans to prohibit by 2015 all products and processes containing or releasing carcinogens, mutagens, endocrine disruptors, and reproductive toxicants, as well as the heavy metals lead, mercury, and cadmium.

Researchers have estimated that the additional cost of REACH will be between $3.5 billion and $6.5 billion over 15 years, but that it will save $60 billion over the same period in reduced health-care costs (World Bank 2010). They also expect that any additional industry costs will be fully offset by profits from new, safer products. As a first estimate, one might assume that North Americans are being burdened with unnecessary health-care costs that are at least equivalent to the net benefits of the more stringent European system. The social consequences of developmental problems in children are likely even more tragic.

The chemicals industry is truly global in scope. For example, the majority of chemicals used in Canada are actually imported into the country. It is increasingly clear that what is needed is a global chemicals-management system, likely building from the strongest international example, which may very well be the European system. Both Canada and the United States need to start the catch-up game very soon for environmental and human-health reasons, but also to get a toehold in the very lucrative international green chemistry marketplace.

Emerging Institutional Issues

From a Canada-US transboundary water perspective, three over-arching institutional trends will have to be very closely watched in the coming decades. The first is the globalization of the world economy, along with trade agreements that may eventually restrict the ability of individual nations to preserve their water resources. The second is a more aggressive approach by state governments in shared basins, which speaks to the phenomenon of rescaled governance detailed in the Introduction and chapter 3 of this volume. Some relevant water examples include Montana's petition to reopen apportionment arrangements for the St. Mary and Milk Rivers (chapter 8, this volume), North Dakota's construction of an outlet from Devils Lake (chapter 9), and Maine's ordering dam operators to close the fish way on the international St. Croix River to the passage of alewives. The third over-arching trend is the decentralization of water management in Canada and the United States, due in part to a significant decline in water management expertise within the federal governments and in part to a conscious decision to create a more distributed capacity.

Conflict and cooperation will always coexist in the Canada-US water relationship. Global issues such as energy insecurity, climate change, exponentially escalating demands for nonrenewable resources, intensifying environmental-health issues, global food shortages, and environmental refugees all threaten to intensify conflict. At the same time, cooperation will become increasingly difficult as governmental budgets become more strained. Under those circumstances, the two countries will have to identify and build on whatever institutional strengths are available. Perhaps the most critical of those strengths is the century-old Boundary Waters Treaty (BWT, or Treaty) and the binational

International Joint Commission (IJC), which was established under the provisions of that treaty (see also chapter 4, this volume).

One measure of the IJC's past success is the fact that out of the 120 cases referred to the commission for advice and resolution, only two have resulted in the commissioners failing to reach consensus and reporting separately to their national governments. Furthermore, in the vast majority of cases, governments have acted on the IJC's recommendations, many of which are explored in Part II of this volume. Although there continue to be local irritants along the border, including those related to inter-basin diversions, cross-border water pollution, and the environmental impacts of mining proposals and cross-border flooding, these are generally cases that have yet to be referred to the commission. Given the very serious challenges lying ahead, it is clearly in the best interest of both Canada and the United States to empower the IJC to work with governments at all levels, as well as with civil society, to capture opportunities to prevent and resolve water disputes. The Treaty is particularly important to Canadians, because the principles embedded within it allow Canada to deal with its much more powerful neighbour as an equal. The treaty continues to be viewed by many as exemplary, and water managers all around the world consider it a useful model. Most Canadian water professionals believe that Canada could not negotiate as favourable a treaty in today's political climate.

Unfortunately, two disturbing trends are hindering the IJC's ability to respond to threats facing boundary and transboundary waters, and its future utility as a vehicle for conflict resolution (Pentland and Sandford 2009). First, in cases like Devils Lake in North Dakota (see chapter 9, this volume), expedient political processes have increasingly taken the place of the sound technical approach offered by the IJC. This is also the case with other water-supply projects in the US Midwest, as well as in the Great Lakes – St. Lawrence Basin, where the IJC's role in overseeing the Great Lakes Water Quality Agreement has been eroded in favour of a more politically driven mechanism. Notably, this approach coincided with a decline in the agreement's effectiveness in addressing pollution in the lakes, as witnessed for example by a return of eutrophication in some areas, as well as a weakening of government accountability and public engagement. This trend, if continued, will invariably work against the best interest of the Canadian government and its people because the country with the least political clout will almost always come out second best. This trend will also generally yield less satisfactory environmental and economic outcomes in both countries because decisions are more likely to be based on incorrect technical assumptions.

The second troubling trend is the decline in Canadian, and to a lesser extent American, scientific capacity in recent years. This decline has further compromised the IJC's ability to protect water resources because the commission relies heavily on credible scientific experts within governments to assist in fact-finding efforts. For example, it has been estimated that governmental capacity in Canada has been reduced by about a third over the past two decades. This reduction has resulted in the development of a serious asymmetry in IJC support in the two countries. Typically, a US Army Corps of Engineers colonel with a staff of thousands and a budget of many millions of dollars might be sitting across the table from a relatively junior Environment Canada engineer with a staff of two or three and almost no financial resources.

Conclusions

As the IJC enters its second century, it faces both some disturbing trends and some very promising opportunities. One of its greatest challenges is the lack of appropriate support from governments, as was pointed out by the Auditor General's Office in 2001 as well as by a Parliamentary Committee in 2004. This lack of support is largely a result of underfunding of the Commission, governments' reluctance to use the Commission to its full potential, and a general decline in support for water and environmental science and policy capacity, especially in Canada. Fortunately, promising signals now from Washington indicate that the US administration is committed to more fully supporting and respecting the independence of bodies like the IJC, and the legitimate contribution of science in decision-making. Assuming that there is follow-through on those promising signals, then there is a high probability that Canadians will eventually follow in US footsteps.

Another way to enhance the IJC's contribution would be for federal, provincial, and state governments to embrace more fully the commission's International Watersheds Initiative, which, as outlined in chapter 4, aims to move toward integrated water-resources management.

As water and related issues along the world's longest undefended border become more complex and interrelated, it will be necessary to engage all stakeholders in a comprehensive way. This will both help increase their understanding of scientific and management approaches, and allow them to voice their perspectives on a desirable future. Federal, provincial, and state governments, together with the IJC, can and must provide top-down policy frameworks and transfer knowledge to citizens in an understandable way. Beyond that, however, the fate of most basins will be determined largely by citizen's initiatives at a fairly local level.

Broader-based watershed institutions could potentially create and facilitate a two-way flow of knowledge, which will be necessary to improve the effectiveness of water-related environmental management at all levels and to help citizens adapt to situations like climate change. Essentially, what is needed in many transboundary basins and will be increasingly needed in the future is an approach that recognizes the complex interplay between socio-economic and environmental factors, quantity and quality concerns, and various segments of society, including Aboriginal peoples (see chapter 2).

Critics sometimes raise the issues of cost and interjurisdictional complexity. However, the incremental costs need not be high since individual issues are inevitably going to have to be dealt with anyway, and a neutral focus like the International Joint Commission could minimize intergovernmental tensions (Pentland 2009).

Finally, if these two countries are to adapt successfully to the geographically diverse and quantitatively serious emerging water issues facing the continent and the world, they will both have to move toward highest common denominator policies and legal regimes. In some cases, those highest common denominator policies and laws can be found in North America. For example, public trust law has played a significant role in US water management in recent decades, and Canada could benefit from something like it. On the other hand, Canadian policies related to inter-basin transfers tend to be more progressive than those in the United States. In yet other cases, such as chemicals management, both countries may have to look beyond North America to find best practices.

References

Bruce, Jim. 2010. "Impacts on Water of a Changing Climate." Presented at the CFCAS Symposium on Canadian Water Security and the Critical Role of Science, Ottawa, 28 May.

Council of Canadian Academies. 2009. "The Sustainable Management of Groundwater in Canada." Expert Panel on Groundwater. Printed in Ottawa, June.

Forum for Leadership on Water. 2010. "Carbon Capture and Storage: Political Folly or Climate Change Fix?" *FLOW Monitor Canadian Water Policy Watch* 2:1–12.

Hewitt, Mark, and Mark Servos. 2001. "An Overview of Substances Present in Canadian Environments Associated with Endocrine Disruption." *Water Quality Research Journal of Canada* 36 (2): 191–213.

Kolpin, D.W., E.T. Furlong, M.T. Meyer, E.M. Thurman, S.D. Zaugg, L.B. Barber, and H.T. Buxton, and the US Geological Survey. 2002. "Pharmaceuticals, Hormones, and Other Organic Wastewater Contaminants in U.S. Streams, 1999–2000: A National Reconnaissance." *Environmental Science & Technology* 36 (6): 1202–11. http://dx.doi.org/10.1021/es011055j. Medline:11944670

Parfitt, Ben. 2010. "Fracture Lines: Will Canada's Water Be Protected in the Rush to Develop Shale Gas?" Munk School of Global Affairs, University of Toronto. http://www.powi.ca/pdfs/groundwater/Fracture%20Lines_English_Oct14Release.pdf.

Pearse, P.H., F. Bertrand, and J.W. MacLaren. 1985. "Currents of Change: Final Report, Inquiry on Federal Water Policy." Ottawa: Environment Canada, 224.Pentland, Ralph. 2009. "The Future of Canada-U.S. Water Relations, the Need for Modernization." *IRPP Policy Options Magazine*, 60–4.

Pentland, Ralph. 2010a. "Coping with Water Quality and Related Ecological Implications of Climate Change." *Contrastes Magazine*, 15 April.

Pentland, Ralph. 2010b. "Do We Have Time to Get Rich First and Get Smart Later?" Presented at the CFCAS Symposium on Canadian Water Security and the Critical Role of Science, Ottawa, 28 May.

Pentland, Ralph. 2010c. "The Changing Face of Risk." Panel following presentation of the Environment Commissioner's report, Munk Centre for International Studies, University of Toronto, 14 December.

Pentland, Ralph, and Adéle Hurley. 2007. "Thirsty Neighbours: A Century of Canada-U.S. Transboundary Water Governance." In *Eau Canada: The Future of Canada's Water*, ed. K. Bakker, 163–82. Vancouver: UBC Press.

Pentland, Ralph, and Bob Sandford. 2009. "The International Joint Commission and the Future of Boundary Water Security." *FLOW Monitor Canadian Water Policy Watch* 1: 1–12.

Quinn, Frank. 2006. "Continental Divides." Presentation to Freshwater for the Future Conference, Policy Research Initiative, Government of Canada, Quebec, 8–10 May. http://www.horizons.gc.ca/doclib/PS_SD_Quinn_200605_e.pdf.

Quinn, Frank. 2007. "Water Diversions, Export and Canada-U.S. Relations: a Brief History." Program on Water Issues, Munk Centre for International Studies, University of Toronto. http://www.munkschool.utoronto.ca/assets/files/Publications/0742QuinnAug-07.pdf

Thompson, Graham. 2009. "Burying Carbon Dioxide in Underground Saline Aquifers: Political Folly or Climate Change Fix?" Munk Centre for International Studies, University of Toronto. http://beta.images.theglobeandmail.com/archive/00242/Munk_Centre_Paper_242701a.pdf

Weibust, Inger. 2009. *Green Leviathan: The Case for a Federal Role in Environmental Policy*. Burlington, VT: Ashgate Publishing.

World Bank. 2010. "International Experience with Toxic Chemicals Management." World Bank Analytical and Advisory Assistance Program. http://siteresources.worldbank.org/INTEAPREGTOPENVIRONMENT/Resources/Chemicals_Int_Experience_EN.pdf

PART TWO

Flashpoints, Conflict, and Cooperation

7 The Columbia River Treaty

JOHN SHURTS AND RICHARD PAISLEY

The United States and Canada are parties to the Columbia River Treaty, first signed in 1961 and ratified in 1964. The treaty obligated Canada to construct three water-storage dams in the portion of the Columbia River Basin in the Canadian province of British Columbia (BC). It then called for the ongoing, coordinated operation of storage and hydroelectric projects in BC and the United States for the dual purposes of flood control and power generation. Both nations understood that most of the power and flood-control benefits of these treaty-storage projects would be realized in the downstream nation (the United States), and that these benefits would then be equitably shared via money and the delivery of power with the upstream nation (Canada), which hosts the projects. The treaty "entities," implementing agencies designated under the treaty, have implemented the treaty for these purposes and under these premises ever since.

The Columbia River Treaty is perhaps the classic example of a successful, benefits-sharing international river treaty, at least when viewed on its own terms. The treaty-storage projects have been an important factor in preventing the damaging flood events that occurred in the United States in earlier periods, such as the devastating flood of 1948 that wiped Vanport, Oregon, off the map.[1] These projects have likewise helped the system operators optimize hydroelectric generation in the region to meet a winter-peak electricity demand. It can always be asked of any particular period in the implementation of the treaty whether the benefits shared back north have equitably matched the benefits gained south, or whether one nation is getting the better end of the deal. But benefits have been shared, and presumably, comfortably so within the bargain's range. Moreover, the operating Entities have cooperated remarkably well in implementing this arrangement. The treaty's success in these terms is probably in part a function of the relative simplicity of the chosen task: flood control and winter-peak power optimization in a hydro-dominated system, with most of the benefits realized in the south and power or money resulting from power generation flowing north to balance the benefits. This task is relatively simple, that is, compared to all of the management issues that bedevil big river basins, international or domestic.

Yet this simplicity is precisely the problem now. The Columbia River Treaty has become an anachronism, in that what both nations wanted in 2011 out of the river and the regional storage and power system, as well as out of the regional decision-making processes, appears to be so very different and more complex than in 1960. However, an opportunity to focus our attention on this dilemma is arising because of provisions in the treaty that (1) automatically put an end to the assured, systematic flood-control operation in 2024 unless the Entities or the nations act to preserve it, and (2) allow for unilateral termination of the power provisions if one or both nations no longer sees the benefit of continued cooperation.

What is needed now is for representatives of the two nations to sit down again and ask themselves the same questions they asked in the 1950s, but in the altered context: Are there ways in which the two nations can continue to cooperate on Columbia River management that would bring greater total benefits than if the two nations act unilaterally? And, if so, what are the right mechanisms for this cooperative operation and to what ends? What are the right mechanisms for estimating and equitably sharing the benefits?

The Columbia River

The Columbia is one of the great rivers of North America. It is the fourth-largest river on this continent, behind only the Mississippi, St. Lawrence, and Mackenzie Rivers. Beginning at Columbia Lake in British Columbia, the main branch of the river travels 1,200 miles through 14 dams before reaching the Pacific Ocean far to the south, 100 miles downstream from Portland, Oregon, in the United States (see Figure 7.1). The Columbia River drains a basin that covers nearly 260,000 square miles (670,000 square kilometres), approximately the size of France, and includes portions of seven US states and British Columbia in Canada. Tributaries feeding the Columbia include several major rivers of the Pacific Northwest, many of them transboundary in nature as well, including: the Kootenai (or Kootenay, as it is spelled in Canada), Flathead, Clark Fork/Pend d'Oreille, Kettle, Okanogan, Methow, Spokane, Wenatchee, Yakima, Snake, Clearwater, Salmon, Owyhee, Grande Ronde, Walla Walla, Umatilla, John Day, Deschutes, Hood, Willamette, Klickitat, Lewis and Cowlitz Rivers.

Figure 7.1 Map of the Columbia River Basin.

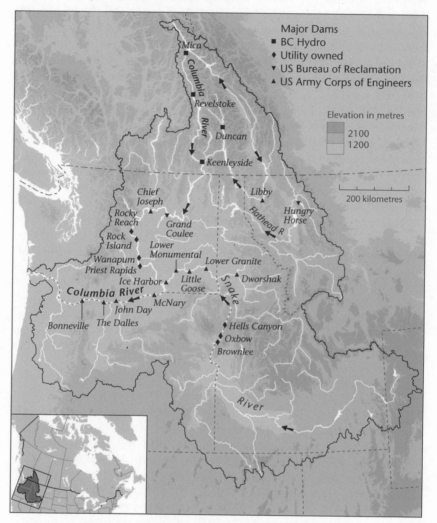

Source: Original Map by Eric Leinberger, University of British Columbia.

The Columbia and its tributaries run through highly variable climatic conditions and topography, from alpine to desert to rainforest. Close to the ocean, yet in a basin laced with tall mountain ranges, the Columbia is relatively unusual in terms of its runoff and the diversity of its habitats. From its headwaters to its mouth, the river drops steadily at a rate of about two feet per mile, and most of its course is through rock-walled canyons. The Columbia pours prodigious volumes of water down these rocky canyons, emptying an annual average of nearly 200 million acre-feet of water into the Pacific. Much of the river's volume originates in its middle and upper reaches, yielding an average annual runoff volume of 134 million acre-feet, measured at The Dalles approximately 200 miles upstream of the mouth. Tributaries west of the High Cascades contribute the rest. The Canadian portion of the basin, which totals 15 per cent of the basin's drainage area, contributes greater than 35 per cent of the average annual flow when measured at The Dalles and about 20 per cent of the river's total volume at the mouth. Most climate-change models indicate that the percentage of the river's total runoff originating in Canada could increase significantly.

Most of the "storage" in the Columbia is natural storage in the form of snowpack in the headwater mountains. Snow runoff has the greatest impact on river flows, with a late spring/early summer big peak hydrograph. The peak flows are obvious: average unregulated fall and winter flows (September – February) measured at The Dalles are around 100,000 cubic feet per second (or 100 kcfs), while May – June unregulated flows at the same point average more than 440 kcfs. The highest peak flood flow at The Dalles of 1,240 kcfs occurred on 6 June 1894, and about half of that flow came from Canada. The variation between years is large, with unregulated peak flows as low as 36 kcfs in 1937, a 1:34 ratio of low to high flows that can be compared to the St. Lawrence's ratio of 1:2 or the Mississippi's ratio of 1:25. Flows at the Canada-US border can range between 14 to 555 kcfs, a 1:40 variation that is even higher than that of the river as a whole.

The Columbia River is home to a number of important species of fish and wildlife, including six species of Pacific salmon: chinook, coho, sockeye, chum, pink salmon, and steelhead. The basin's salmon and steelhead runs were once among the largest in the world, with an estimated average of 10 to 16 million adult fish returning to the basin annually. For thousands of years, the Indigenous people of the basin have depended on these salmon runs and other native fish for physical, spiritual, and cultural sustenance. Commercial and sport fishing, as well as recreational, aesthetic, and cultural considerations, endear fish species such as salmon and steelhead to millions of residents and visitors. Many even refer to the salmon as the Northwest's iconic symbol. A number of animals including bald eagles, osprey, and bears also rely on fish from the Columbia River and its tributaries to survive and to feed their young.

Salmon and steelhead runs, along with other native fish and wildlife in the basin, have declined significantly in the last 150 years. Recent years have seen a combined total of little more than a million upriver adult salmon and steelhead passing Bonneville Dam, the lowest dam in the river, and many of these fish came from hatcheries. Overfishing certainly contributed to this decline, but land and water developments across the region also played a role, blocking traditional habitats and dramatically changing natural conditions in rivers where anadromous and resident fish evolved.

These developments included the construction of dams throughout the basin for such purposes as hydroelectric power, flood control, commercial navigation, irrigation, and recreation. The combination of high volume and stable canyons made the Columbia an ideal river for storage and run-of-the-river hydropower dams. Today, 14 dams are on the mainstem of the Columbia, beginning with Bonneville Dam and ending with Mica Dam in British Columbia 1,000 miles upriver. A dozen more can be found on the mainstem Snake River, the Columbia's longest tributary at 1,000 miles, as well as large storage or power-producing dams on several other tributaries such as the Kootenai, Flathead, Pend d'Oreille, Clearwater, and Willamette. In total, more than 450 dams exist throughout the basin, making the Columbia one of the most dammed rivers in the world, especially for hydropower purposes.

Water storage in the Columbia River totals approximately 30 per cent of the average annual runoff, which fluctuates from year to year depending on the snowpack. The basin produces, under normal precipitation, about half (16,200 average megawatts) of the electricity consumed in the Pacific Northwest, an unusually high percentage compared to anywhere else in North America. Dams clearly affect how water flows in the modern Columbia River by: storing runoff, reducing flood flows, shifting flows from the natural spring/summer peak to fall/winter to meet the region's peak electricity demand, slowing river velocities at run-of-the-river projects, and blocking, inundating, or reconfiguring major river reaches. These river developments support the region's economic prosperity, but have substantial adverse effects on the basin's fish and wildlife, the river conditions that support this biological diversity, and the people who value those fish, wildlife, and river amenities.

Columbia River Treaty Particulars

In the middle of the last century, many people living in the Columbia Basin, especially in the cities and farming areas south of the border, saw the Columbia's big spring peak runoff as naught but a problem and possibly an opportunity. Peak flows in higher runoff years brought the river out of its banks at various points, presenting an obvious obstacle to efficient development of cities and farmland, especially troublesome in the more developed areas of the lower river. This problem came to a head on 31 May 1948 when a huge flood that damaged homes, farms, and levees from British Columbia to Astoria caused more than $100 million in damage in 1940s dollars. The flood destroyed Vanport, Oregon, a city of 35,000 on the river near Portland, killing more than 50 people. The US Geological Survey estimated that the 1948 flood flow was over 1 million cfs, and the river crested 25 feet above normal at Vancouver, a city directly across from Vanport. Flows at Grand Coulee Dam, 600 miles inland, peaked at 633 kcfs that spring, more than three times the average at that location during that time of the year.

In addition to the flood problem, people also thought hard in mid-century about the unrealized potential of hydropower generation on the river. Hydroelectric dams on the US portion of the river began powering the developing economy of the region in the 1930s and 1940s, with plans that later came to fruition to add far more hydropower generation after the Second World War. Hydroelectric power became the dominant form of electricity generation in the Pacific Northwest, as nowhere else in North America. The relatively cool Northwest developed a winter peak in electricity use, and yet about three-quarters of the annual average flow of the Columbia River ran in the late spring and early summer due to the melting snowpack. With the major exception of Grand Coulee, the Columbia hydro projects built in the United States were largely run-of-the-river, meaning they stored little water. When it wasn't covering their lands, a lot of the water ran "wasted to the sea," and certainly ran at the wrong time of year in terms of matching electricity generation to peak-power demands.

With these two issues in mind, it was not lost on some that most of the best reservoir storage sites were in Canada, where much of the runoff could be captured and managed. Constructing storage projects at these sites to contain a significant part of the snowpack runoff would help decrease flood risks by reducing the big peak flows, particularly protecting US lands from floods. It would also help optimize the generation of electricity, again especially at the US projects. The stored water could be released in the winter, augmenting the lower base flows and thus increasing the amount of electricity generated at the hydro projects in the lower river, which would better match the Northwest winter peak demand.

These realities, however, created an interesting dynamic for negotiations between the two countries. Major storage projects would be in Canada, which means that this is also where the costs and burdens would fall, not only in terms of money but also inundated lands and dislocated land uses and communities. Meanwhile, the expected benefits at that time, and as projected into the future, would largely occur in the United States, including significant flood protection, increased generation from the US hydro projects, and the bulk of electricity demand and use. So, how might Canada be induced to build the projects? How might the United States be induced to share the costs and benefits?

The Columbia River Treaty was the eventual result. Discussed, analysed, and then negotiated through the late 1940s and 1950s, the United States and Canada agreed to the treaty terms in 1961. Ratification, however, did not take place until 1964. This lag was due to the need to work out a separate arrangement in Canada between the national government and the province of British Columbia. The national government largely transferred Canada's rights and obligations under the treaty to the province and authorized a sale by the province of its Entitlement to the downstream power benefits (as discussed below). The province was then to add a protocol to the treaty to memorialize this arrangement and clarify other matters. The specific treaty on how to develop and manage the Columbia superseded the more general Boundary Waters Treaty earlier in the century, meaning the International Joint Commission has no role in Columbia River Treaty matters (for a detailed discussion of the IJC, see box 1.1 and chapter 4, this volume).

The basic provisions of the treaty are relatively simple to explain. Canada agreed to build three dams in the Columbia Basin, providing 15.5 million acre feet (maf) of storage: Keenleyside (at Arrow Lakes), Mica on the mainstem Columbia, and Duncan on an arm of Kootenay Lake. Canada then agreed to dedicate that storage to an operation that would alter river flows to allow for greater winter generation at the US projects. Canada also agreed to operate 8.5 maf of that total storage for the next 60 years (that is, until 2024) in an assured flood control operation. The power provisions do not have an automatic sunset date. Instead, either nation may unilaterally terminate the treaty provisions beginning after 60 years (in 2024), but only after at least 10 years' notice (so at least by 2014).

In return for the significant flood control and power benefits, the United States agreed to pay Canada nearly $65 million in advance, representing half of the estimated value of the reduction in flood damage in the United States through 2024. The United States also agreed to deliver half of the "downstream power benefits" the treaty provided (that is, the increment of additional value in generation at the United States projects resulting from the coordinated Canadian storage operation) to British Columbia, a benefit also known as the Canadian Entitlement. The two nations chose to focus on sharing gross benefits through an equal division of the power generated, rather than attempting to calculate the precise net benefits realized in each country. This latter strategy would have required trying to quantify and balance all possible costs and opportunities related to the new storage dams, a nearly impossible task. This insight is important to keep firmly in mind as the region reshapes the treaty arrangement to express additional values and purposes.

The treaty required each country to designate an Entity for implementation. Canada's selection is actually a British Columbian Entity: BC Hydro, a provincial corporation that generates and sells most of the electricity in BC from a host of province-owned hydro-generation units on the Peace, Columbia, and other rivers. The US Entity is shared between the administrator of the Bonneville Power Administration (marketer of power from US dams on the Columbia) and the division engineer of the Northwest Division of the US Army Corps of Engineers (operator of most of the federal Columbia dams and responsible for implementing system-wide flood control). The Entities have each appointed treaty coordinators and a secretary, and established a joint treaty-operating committee from agency personnel to oversee annual planning and operations under the treaty.

The Columbia River Treaty sets forth a fairly complex planning process for the Entities to follow. The treaty requires that each year, the Entities develop an Assured Operating Plan (or AOP) for the treaty projects from the sixth succeeding operating year. Nonpower and nonflood control considerations cannot be considered or included in an Assured Operating Plan. At the time of the Assured Operating Plan, the Entities also determine the downstream power benefits, and thus, the Canadian Entitlement, that would result from those operations in that sixth year. The downstream power benefits are determined under a planning assumption that the United States will optimize the operation of its power generation projects in that year. For example, in November 2004, the Entities agreed to the Assured Operating Plan and determined the downstream power benefits for the 2009–2010 operating year. Actual operations in that future year do not affect the determination of the downstream power benefits and the delivery of the Canadian Entitlement – those benefits are set in the plan six years ahead of time and are not to be changed.

The Entities then also agree to a Detailed Operating Plan each year just in advance of that operating year. For example, the Entities agreed in July 2009 to the Detailed Operating Plan for the 2009–2010 operating year, which began in August 2009. The Detailed Operating Plan is, as its name implies, a detailed plan of operations based on more extensive and up-to-date studies than in the Assured Operating Plan from five years before. The Entities may agree to a Detailed Operating Plan that produces treaty operations and results that are "more advantageous" to the parties than would result from operations under the Assured Operating Plan. As noted above, however, the treaty does not allow the Entities to recalculate the downstream power benefits and Canadian Entitlement at the time of the Detailed Operating Plan or after the fact of actual operations. Those benefits remain as determined in the Assured Operating Plan from six years before.

The treaty also authorized the United States to construct Libby Dam on the Kootenai River, with a storage pool of another five maf backed up into British Columbia. Outflows from Libby are part of the Kootenai River flows that also return to Canada. Libby is not operated as part of the treaty, but its operations must be coordinated with the treaty project planning and operations. The benefits of this dam also do not factor into the calculation of the downstream power benefits and the Canadian Entitlement; each nation simply receives whatever benefits may be realized from flow regulation at Libby.

The Challenge of Anachronicity

For nearly 50 years, the Entities have implemented the Columbia River Treaty and operated the projects under these premises. As was mentioned in the introduction, the treaty has proven to be a well-conceived and effective benefits-sharing treaty, on these premises. The current problem is that the treaty is also an anachronism, "a thing that is chronologically out of place; especially a thing from a former age that is incongruous in the present." The treaty has developed this quality because what both nations wanted in 2011 out of the river, the regional power and storage system, and out of regional decision-making processes seems so different from what was wanted in 1960. The following section highlights some of the key elements of this dilemma. Two elements are particularly notable by their absence, given the developments of the last 50 years – (1) the absence of ecosystem considerations in treaty purposes or the formal planning for treaty operations, when ecosystem considerations are now integrated into the policy and law of the river on both sides of the border, and (2) in a Columbia Basin in which the multiple sovereigns and the public are critical participants in decisions about the river on both sides of the border, the complete absence of tribal, state, regional, and public input or participation in the planning and operation of the treaty projects.

From the vantage point of 2011, the most obvious substantive elements missing from the treaty are fish and wildlife, water quality, and other environmental and ecosystem benefits. These are not included in treaty purposes, planning, the calculation of benefits, or formal treaty operations. Yet by the twenty-first century, these considerations have become critically important elements of river law, policy, and values on both sides of the border. Such considerations have also been woven into international laws and arrangements on transboundary rivers across the world. In the United States, fish and wildlife protection, mitigation, and species conservation, especially for salmon and steelhead, have become major drivers of system management in the Columbia, on an equitable basis with other project purposes. The federal agencies plan and operate the big storage projects at Grand Coulee Dam and Libby Dam, for example, with fish and wildlife considerations woven as deeply into the fundamental management concerns as any other purpose. The US government also spends hundreds of millions of dollars per year to mitigate the system's impacts on salmon and steelhead populations, as well as other fish and wildlife. British Columbia has preferred operations to protect local spawning and rearing conditions for whitefish and trout, in addition to fish and wildlife mitigation programs all along the river especially in and around Kootenay Lake. Nonetheless, the joint-management scheme for the Columbia River in its international dimension is silent on these values. It is hard to imagine that this formal neglect can continue.

Even if optimum power generation was the only benefit that the two nations still cared about, the treaty-power operation may also be a relic from another age. The regional power system is not only fairly different now than it was in 1960, it is significantly different in a way other than was projected in 1960. The Columbia River Treaty negotiators assumed that the regional-power system would significantly increase the number of baseload coal and nuclear thermal plants over time, reducing the relative value of hydro-generation. This did not happen. The negotiators did not foresee the following: the region's considerable investment in conservation; the addition of significant megawatts of capacity in relatively small natural-gas plants that are best used for peak loads; the recent addition of large amounts of intermittent wind power, which cannot be factored into planning, and the assessment of benefits to the system under the treaty provisions; the development of a large-scale, wholesale power market across the west, including increasingly valuable markets in the summer; or the fact that the benefits to BC Hydro and the province from generation at the treaty projects (and at Revelstoke Dam below Mica), which are also not factored into the benefits calculation under the treaty, would become so much more valuable than the Canadian Entitlement. These developments have contributed to the continued increase in the value of the river's hydro-generation, contrary to expectations in 1960. They have also complicated the management and integration of resources with hydro-system operations, and presented new challenges and opportunities at different times of the year for optimizing power generation and generating revenue. Power-system optimization, as required under the treaty-planning protocols, no longer matches current power-system operations or what might be the optimal power-system arrangement in the future.

The one element of the treaty operations that may not be anachronistic – the use of the treaty storage to provide assured flood risk reduction in the United States – is ironically the one provision that ends automatically in 2024! The only anachronism here may simply be that the United States did not have the foresight to pay for more than 60 years of its share of flood-control benefits, and this amount was woefully under the mark of the actual benefits realized. What will be left in its place is the concept of "called-upon" flood control: the United States may still call upon the Canadians to use the projects to provide flood control, but only under certain imposed conditions and costs, and with almost no clarity about the details, which will certainly matter. The two things that do seem obvious with a "called upon" operation is that if the United States wants the same level of flood protection, (1) the US storage projects will have to bear a significantly increased share of the flood-risk reduction, to the increased detriment of fish and other biological qualities in the reservoirs and downstream of these projects, and (2) the United States will pay significant new costs.

The Entities are not completely paralyzed by the anachronistic context even now. Each year, they execute what are known as annual supplemental operating agreements in what can appear to be an increasingly strained effort to realign actual project operations from what is first required under the treaty to something that more closely relates to current needs and opportunities concerning flows for power and fish. The Entities do this under the guise of a somewhat ambiguous treaty provision, which allows them to prepare detailed operating plans for the coming year "that may produce results more advantageous to both countries." The Entities can interpret this phrase to allow for the integration of new and different purposes for operations. But the obvious question is whether these annual supplemental agreements are sufficient in incorporating an entirely new set of needs and considerations in a robust and equitable way. Can these annual ad hoc agreements continue to bear that weight in addition to taking on the added weight of further-evolving power operations, the replacement of "assured" flood control with the messiness of "called upon" flood control, and the effects of climate change? Does it not make sense to build these other purposes and needs more formally into the firmament of the international relationship itself? At the same time, it is possible to imagine the negative consequences of trying to do this. These include the difficulties of amending the treaty and the possibility of losing regional control to national negotiations. The negative consequences may also include the loss of a smoothly functioning technocratic treaty operation when we try to balance competing values and needs that are much harder to determine, quantify, and integrate in a mechanical way, as compared to flood control and power generation alone.

Moreover, one important element of the treaty may not be amenable to realignment through an annual supplemental operating agreement: the way in which the benefits of treaty operations are calculated and then shared. As noted above, the treaty requires that the downstream power benefits be calculated on a six-year-ahead planning basis, a process that allows the Entities to calculate benefits derived only from the classic power optimization principles that formed the treaty in the first place. The United States no longer operates its projects to optimize for power generation. Major factors in US project operations now include legally required nonpower constraints governing the use of storage and spillways for the protection of salmon and steelhead and sturgeon, with significant on system generation. Yet this change cannot be considered in the benefits calculation under the treaty, nor can the fact that the actual treaty project operations under the supplemental operating agreements will differ from the classic planned operations. The value of the system as it relates to newly added wind-power generation or new market opportunities also cannot be factored in. That was the deal made in the 1960s, and so these two nations live with the deal. This is not to say that the amount of benefits currently delivered is obviously incorrect, unfair, or out of balance, but the method by which power benefits are calculated and shared has become completely hypothetical, not aligning well with operational or domestic legal reality. Is the way in which treaty benefits are currently determined and shared logical? Is it equitable? Is there just a problem here, or also an opportunity? If it is a problem to resolve, how can the Entities and the region shift the way in which benefits are calculated and paid while ensuring that benefits are still equitably shared? Is it at all possible to do this without having to amend the treaty substantially?

A final anachronism is the nature of decision-making under the treaty. As discussed elsewhere in this volume, public review and participation has become a hugely important part of natural-resource management by governments and agencies on both sides of the border since 1960 (see especially chapter 3). The same is true, at least in the United States, with regard to significant participation by the other sovereign entities in the basin – the states and Indian tribes – in the river-management decisions of the federal agencies operating the projects or engaged in regulatory or mitigation activities. This inclusive participation is not part of the treaty, with the obvious exception of British Columbia and BC Hydro as the Canadian treaty partner. Planning and operations under the Columbia River Treaty take place without public review or participation, and without review, participation, or input from state governments, tribes, or First Nations. Given the current domestic expectations for public participation in river decisions and for sovereign participation in a federalism arrangement, there is no way that the United States and Canada will be able to modify, evolve, or terminate this treaty without building in significant sovereign and public involvement in some way. What can be done to satisfy this need while preserving an international operating agreement between two nations? How do we do so without losing one of the great benefits of current treaty operations, which is how easily and smoothly the Entities operate year-to-year and day-to-day without public interference?

The Opportunities Inherent in "Termination Chicken" and Imposed Transformation

A recent opportunity to focus our attention on this dilemma has arisen, in particular because of the logic of the 2014/2024 termination clause in the Columbia River Treaty. This is so even though the possibly tumultuous end of the assured flood-control operation in 2024 may ultimately serve as the major driver in inducing action for change. This is also true even though the two nations could mutually agree to deal with these problems in a different way at any time. They do not need the game of "treaty termination chicken" to have this discussion, but that may be what it takes. Termination itself has some attractions and some obvious downsides, and would certainly be an unusual event in the law of international rivers. In any event, the Entities have kicked off what they are calling the 2014/2024 Review, and we are well into that review.

What we see as needed now is for representatives of the two nations to sit down again and ask themselves the same question they asked in the 1950s, but in the modern context. This context is different from in the 1950s both in the sense that domestic values and needs have significantly evolved, and in the sense that the projects are now in place as they were not in 1950s. However, the basic question is the same: are there ways in which the two nations can cooperate on Columbia River management that would bring greater total benefits than if the two nations act unilaterally? If so, what are the right mechanisms for this cooperative operation and for estimating and sharing equitably the benefits? Alternatively, what purposes, needs, or values might be better addressed by each country acting unilaterally, taking into account the effects of cooperative treaty operations for a limited set of purposes?

If and when we ask ourselves these questions, we also need to consider the fact that the future is uncertain and grapple with that uncertainty, as compared to the certainty with which the present treaty incorrectly anticipated the future. How do we introduce adaptive flexibility into the treaty while providing the necessary certainty at any one time? This will only become more important in light of climate change, which will certainly increase the level of uncertainty. The conversation will also have to include whether it is possible to reshape current treaty operations and the ways in which benefits are shared along these lines without having to go through the painful process of a formal treaty renegotiation.

If representatives of the two nations have this conversation, it is likely to involve Entity representatives primarily, at least in the first instance and in the primary sense. If so, then how are other voices, both public and sovereign, heard? This needs to be a regional conversation, broadly involving the people and communities of the river.

A Modest Proposal

Thus, before the dust settles, the Entities, the Pacific Northwest region, and the two nations will probably find it necessary or desirable to revise their current arrangement after serious consideration of (1) ecosystem, fish, and wildlife considerations as an equal value on the river; (2) the role of the treaty projects in an evolved regional-power system; (3) the need for a sensible way to provide assured flood control; (4) the needs of those whose lives and livelihoods depend on the recreational use of upriver reservoirs; (5) a new way to determine and share the benefits of project operations that bears some connection to current reality; (6) demands for broader public and sovereign participation; (7) the balance between adaptive management flexibility, and planning and operational certainty; and (8) a vehicle for making the transition, ideally one that allows the region some semblance of control in resolving these issues and does not hold them hostage to unrelated national interests and issues.

We wish them luck.

Actually, we are optimistic it will be done. The two national governments and the people in the region have a long record of successful cooperation in delivering and sharing the benefits of the Columbia River. We believe that the United States and Canada will draw on that tradition and emerge from the morass into a reformed cooperative arrangement that we will be proud to own.

We also assume that there are many ways to reach that end. One suggestion developed here is a realistic scenario to improve ecosystem conditions from the headwaters to the estuary by following an operational profile for all storage projects (in both the United States and Canada) that is more consistent with natural hydrographic patterns than current treaty-project operations. This change would have to be balanced with the continuing provision of an important level of system flood control and will also provide continuing, if somewhat different, power-generation benefits and summer recreation amenities. One key principle to this scenario is that the benefits of this operation should continue to be shared equitably. This might take place through the continued delivery of generated power or a share of hydropower revenues, although calculated in a different way, or through other mechanisms. This scenario also provides for both public and sovereign participation in a reasonable way, as well as for adaptive management flexibility over time while preserving a stable planning and operational regime. We believe that this can be accomplished without needing to amend the treaty. This is "pie in the sky" optimism, perhaps. But, a more detailed sketch of this scenario follows.

Operations

Over the last decade or so, the United States has changed the operation of its storage projects with the intent to create improved ecosystem conditions for fish, wildlife, and water quality. A management purpose equal to providing flood-control benefits and power generation, this has come to mean a particular operating protocol for the US storage reservoirs. The suggestion here is to apply this same or a similar operation to all the main storage projects in the basin on both sides of the border, including the Columbia River Treaty projects. The key elements of this operation would be to:

- Operate all of the main US and Canadian storage projects in a similar fashion as a system.
- Spread an assured flood risk-protection operation across all the storage projects, with a system aim to manage flows so that they do not exceed the upper range of protection that is currently managed for at The Dalles. This management level will leave somewhat greater flood risk than currently achieved under the treaty operations. The United States and Canada will then have to achieve any additional flood risk management through local and nonsystem measures, including considerations of nonstructural alternatives. The eventual method for calculating and equitably sharing benefits will have to acknowledge that Canadians are extending assured flood control for the indeterminate future to the great benefit of the United States.
- As an operational priority, keep the storage-reservoir levels as high as possible coming out of winter and into spring to pass as much of the spring/early summer runoff as possible. Storage projects should be no lower than their upper flood-control limits by mid-April. Any power generation drafts in winter that draw projects lower than the flood-control limit should be replaced before then. This would be the most significant change from current treaty-project operations, and the purpose would be to set up the right conditions for the following spring operation.
- Pass as much of the runoff as possible in the spring, consistent with steady progress toward reservoir refill. Additional storage releases may be decided upon from particular projects for certain localized needs (such as sturgeon flows below Libby, or whitefish and trout flows below Arrow), but the main point will be to minimize reservoir manipulation in return for consistently higher runoff flows due to the winter/early spring operational change. Power generation at run-of-the-river dams will be based on the resulting

river-flow profile, with generation also affected by the continued use of bypass spill needed to improve fish passage through the dams. The purpose of this operation will be to mimic, as much as possible, the natural spring runoff pattern essential to river processes and the conditions that key local species evolved in, consistent with preserving a significant amount of system flood-risk management.

- Place a high priority on achieving refill of the storage projects by their usual target date for refill (end of June, end of July, and so on).
- Ensure that for summer and early fall operations, reservoir outflows largely match inflows. Limited summer-flow augmentation from storage can occur to improve flow conditions below reservoirs and in the lower river for fisheries and generation purposes, but with two important constraints: (1) end-of-summer draft limits that minimize the total summer draft and preserve reservoir levels for biological (and recreational) benefits, and (2) outflows should be held steady or steadily declining throughout the summer, not ramped up or down. The idea is to manage flow more like a headwaters lake, and less like a bathtub.
- Generate month-to-month power-generation patterns as a byproduct of these operations. To the extent that the region's hydropower resource is reduced or reshaped by this operation, US and BC agencies and utilities will compensate by adding efficiency resources and noncarbon or low-carbon-generating resources through their traditional resource-planning methods.

Benefits

Representatives from the two national governments should agree at the start of discussions that the goal will continue to be equitable sharing of benefits across the international border. This is likely to require the continued use of some mechanism to transfer benefits from one nation to the other, to the extent that the burdens and benefits of this revised operation do not affect the two nations equally. There are some key points to consider:

- Canada's balance will be in deficit if BC Hydro continues to provide an assured flood-control operation after 2024. This deficit will have to be made up, either through payments or some form of compensating benefits out of the operation, or some combination of the two.
- Since the power system is so different now than it was in 1960, as described above, it will take some time to analyse the relative power-system benefits

of this revised operation. If the power-system benefits are not balanced, this too can be factored into the benefit/burden accounts of each nation in coming to a final resolution about sharing the overall benefits equitably.

- Assigning quantitative values to the ecosystem benefits of the revised operation is a more problematic task. It may be better to think, in the broadest possible terms, about how the two nations would benefit from improved ecosystem conditions throughout the system. The next step would be to develop a qualitative way to assess the relative proportion of these benefits distributed to each nation and develop a corresponding way to adjust the overall benefits/burdens accounts of each nation. One strong suggestion would be not to try to settle this relationship with finality, but instead, to build in an opportunity to revisit it periodically.

- The two nations should consider using international mediation assistance. Independent legal and technical advice could be useful for assessing the overall benefits of the revised operation, as well as estimating and comparing the proportionate share of the benefits that will accrue to each nation.

- It seems likely that we will still need to share the benefits equitably that are realized disproportionately in one country over the other. We currently use hydropower deliveries from the United States to Canada as a mechanism to make up for the unequally realized power-generation benefits. There is no reason that this mechanism cannot continue to be used, if needed, to share power-generation benefits equitably within the reshaped operation and transformed power system. The two countries might possibly agree to use a different method for determining the total power benefits from current operations. Although perhaps more controversial, the two nations might consider using this mechanism to equitably balance the overall benefits account, to the extent that it seems out of balance after looking at all of the elements. This is a convenient, well understood, valuable, and fungible way to share benefits equitably. The US side, in particular, would need to come to some internal arrangements to make this work, since ratepayer revenues could be seen as being used to compensate for nonpower benefits.

- As noted above, the current method for calculating downstream power benefits is now unrelated to real events and conditions, and will be even more so upon the implementation of this scenario. Yet we might find, after much analytical exercise, that the current method continues to work as a useful surrogate for sharing equitably the overall benefits. If so, it might make sense, upon reflection, simply to continue using the method. This would have the added benefit of making it easier to execute and implement the revised operations via some sort of new annex, protocol, or multi-year supplement agreement to the existing treaty.

Implementation

We suggest the retention of the basic implementation and operating principles, procedures, and institutions, reshaped as needed to implement the revised operating protocol. In addition, we recommend that the Entities look to other international agreements as possible models for how to supplement the current structure with a broader array of sovereign, technical, management, and public perspectives that will better assist in planning, implementation, operation, and assessment. On each side, for example, the respective Entity could establish a formal, small advisory panel of state, tribal, and possibly nongovernmental representatives, with technical assistance as well as broader mechanisms for periodic public input to the advisory panels. The Entities should also consider establishing a joint or transboundary advisory panel drawn from the same spectrum, with independent technical assistance. We suggest the Entities establish these advisory panels and input mechanisms soon, to allow for an advisory role in shaping the revised plan for operations and sharing benefits, as well as in implementation after agreement. In its ongoing 2014/2024 Treaty Review, the US agencies have created a Sovereign Review Team and corresponding technical teams involving state and tribal representatives along with federal. It is a good beginning.

Form of Agreement

Many elements of the operating scenario described here already take place under annual supplemental operating agreements, but not to the extent that would be sufficient for a revision of the magnitude described here. However, we hope it will not be necessary to amend the treaty to accomplish these purposes, which would trigger the complex, national, treaty-approval processes in Canada and the United States. Instead, we should work as hard as possible to develop and agree to a multi-year operating protocol or annex to the current treaty, and adjust the operational elements as necessary without significant compromise to the basic premises described above. This agreement under the treaty should contain the elements described above, including: agreed-upon operation; the details of any changes necessary in the current planning, implementation, and operating procedures; and a description about how the benefits have been or will be assessed and equalized, within the parameters of the existing Columbia River Treaty.

Disclaimer

John Shurts is the General Counsel of the Northwest Power and Conservation Council in Portland, Oregon. Nothing in this chapter reflects the views or positions of the Council or of individual Council members or the staff. Nor should any of it be attributed to Shurts in his official capacity as legal adviser to the Council.

NOTE

1 What many people took from the Vanport event was that society needed to invest some of its wealth in upriver storage to prevent overbank flood flows in major metropolitan and agricultural areas, flooding some lands to prevent the flooding of others. And they acted on that lesson, spurred by the Vanport flood to seriously examine the development of flood-control storage in Canada. We are wrestling with that legacy today, the obvious benefits and obvious and not so obvious costs. Also note that while the focus of the reporting on the 1948 flood has been on the destruction of Vanport, as our summary indicates the river flooded and damaged well beyond Vanport in June 1948 – a famous picture shows the railroad depot and other parts of downtown Portland under water.

REFERENCES AND FURTHER READING

Bankes, Nigel. 1996. "The Columbia Basin and the Columbia River Treaty: Canadian Perspectives in the 1990s." *Northwest Water Law and Policy Project,* Publication No. PO95-4.

Canadian Departments of External Affairs and National Resources. 1964. *The Columbia River Treaty, Protocol and Related Documents.* Ottawa, February. http://www.em.gov.bc.ca/ EAED/EPB/Documents/1964_treaty_and_protocol.pdf.

Columbia River Treaty Entities. 2009a. "Columbia River Treaty Hydroelectric Operating Plan: Assured Operating Plan for Operating Year 2013–14 and Determination of Downstream Power Benefits for the Assured Operating Plan for Operating Year 2013–14." February. http://www.crt2014-2024review.gov/Files/ AOPDDPB14.pdf.

Columbia River Treaty Entities. 2009b. "Detailed Operating Plan for Columbia River Treaty Storage 1 August 2009 through 31 July 2010," July. http://www.nwd-wc.usace.army.mil/PB/PEB_08/docs/dop/ 11DOP.pdf.

Columbia River Treaty Operating Committee. 2008. *Agreement on Operation of Treaty Storage for Nonpower Uses for 15 December 2008 through 31 July 2009,* November.

Governments of the United States and Canada. 1961. *Treaty between the United States of America and Canada Relating to Cooperative Development of the Water Resources of the Columbia River Basin* (Columbia River Treaty), signed 17 January, ratified 1964.

Krutilla, John. 1967. *The Columbia River Treaty: The Economics of an International River Basin Development.* Baltimore: The Johns Hopkins Press, for Resources for the Future.

Muckleston, Keith. 1980."International Management of the Columbia." In *Conflicts Over the Columbia River*, proceedings of a seminar conducted by the Water Resources Research Institute, Oregon State University, July.

Shurts, John. 2009. "Rethinking the Columbia River Treaty." In *The Columbia River Treaty Revisited: Transboundary River Governance in the Face of Uncertainty*, ed. Barbara Cosens and the Universities Consortium on Columbia River Governance. Corvallis: Oregon State University Press.

Swainson, Neil. 1979. *Conflict Over the Columbia: The Canadian Background to an Historic Treaty*. Montreal: McGill-Queen's University Press.

Swainson, Neil. 1986. "The Columbia River Treaty: Where Do We Go from Here?" *Natural Resources Journal* 26: 243.

U.S. Army Corps of Engineers and Bonneville Power Administration. 2009. "Columbia River Treaty 2014/2024 Review: Phase 1 Technical Studies." April. http://www.crt2014-2024review.gov/TechnicalStudies.aspx.

U.S. Army Corps of Engineers and Bonneville Power Administration. 2011. "Columbia River Treaty: 2014/2024 Review." http://www.crt2014-2024review.gov/.

U.S. Army Corps of Engineers, Northwestern Division. 2008. "Columbia River Treaty Permanent Engineering Board."

Volkman, John. 1997. "A River in Common: The Columbia River, the Salmon Ecosystem, and Water Policy." Report to the Western Water Policy Review Advisory Commission. University of Oregon Libraries.

8 Apportionment of the St. Mary and Milk Rivers

NIGEL BANKES AND ELIZABETH BOURGET

The contrast between the rugged mountain setting from which the St. Mary River stems and the arid plain through which the Milk River flows can be startling. Originating in the snowpack and glaciers of the Rocky Mountains, the St. Mary River has an average annual flow nearly five times greater than that of the Milk River. In addition to being smaller, the Milk River's flow is also twice as variable. Settlers in the late 1800s in both the United States and Canada therefore sought to tap the more plentiful and reliable waters of the St. Mary River for irrigation and began drawing up plans, securing rights, and making investments. However, the irrigation schemes that were developed on either side of the international border competed with each other, threatening desired progress and creating uncertainty and conflict. The Boundary Waters Treaty of 1909 (BWT, or the Treaty) contains a unique provision that establishes general procedures to allocate the waters of the St. Mary and Milk Rivers between the two countries, but implementation led almost immediately to further questions. The International Joint Commission's Order of 1921 (the Order), which is still in place today, provides more detail about how to allocate these waters, but questions still arise regarding apportionment pursuant to the treaty and order. Irrigators on both sides of the border continue to depend on water from these two shared rivers, but face many challenges due to the aging infrastructure of the St. Mary Canal, the effects of climate change, and evolving social expectations. Although First Nations and tribes were not involved in the development of the Treaty or 1921 Order, recent institutional frameworks created to examine how to optimize the existing arrangements have included avenues for tribal participation on the US side.

Geographic Overview

The Milk and St. Mary Rivers drain northern areas in the US state of Montana and southern parts of the Canadian Provinces of Alberta and Saskatchewan.

The St. Mary River's water source is in Glacier National Park, Montana, and is part of the South Saskatchewan Basin. The Milk River is part of the Missouri Basin and exhibits the annual and seasonal variability typical of prairie streams. The region is semi-arid, and irrigation is essential for most forms of agriculture.

The St. Mary River, along with the Waterton and Belly Rivers, is one of the three important transboundary rivers that originate principally within Glacier National Park in the United States and flow north into Canada as part of the South Saskatchewan drainage basin. For all three rivers, the largest part of the drainage basin lies within Canada, but a disproportionately larger share of the water supply comes from the part of the basin within Montana. The average annual natural flow of the St. Mary River at the border is estimated at 650,000 acre-feet (AF) (US Bureau of Reclamation 2004, 9). A number of other streams, notably Lee Creek, cross the boundary independently before joining the St. Mary River in Canada.

The Milk River rises nearby in northwestern Montana, but is cut off from mountain water supplies by a low divide (Simonds 1999). The river begins as three separate branches that later join together: the North Fork, Middle Fork, and South Fork. After crossing from the United States into Canada, the Milk River flows eastward approximately 100 miles in Canada before re-entering the United States (International St. Mary – Milk Rivers Administrative Measures Task Force 2006, 12). A number of creeks and coulees join the Milk River as it passes through Canada, but none of them are perennial. Milk River flows are characterized by large annual variability, with most flow attributable to snow melt and sudden large rainfall events. The Milk River's average annual natural flow as it crosses back into the United States at its eastern crossing is approximately 126,900 AF (International St. Mary – Milk Rivers Administrative Measures Task Force 2006, 14). See Figures 8.1 and 8.2.

From its eastern crossing of the border, the Milk River then flows south and east, ultimately joining up with the Missouri River. Tributaries originating in Canada, principally in Saskatchewan, in this portion of the basin are called the eastern tributaries,[1] which flow south into the United States and join the Milk River, mostly downstream of the Fresno Reservoir, the major water storage location on this river.

The Blackfeet, Rocky Boy's, Fort Belknap, and Fort Peck Indian Reservations are located in and adjoining the St. Mary and Milk Rivers watershed. The drainage areas[2] of the South Fork and North Fork Milk River within the United States lie almost wholly within the Blackfeet Indian Reservation. On the Canadian side of the border, both the Kainai (Blood) and Piikani (Peigan) reserves fall within the area generally able to be irrigated by water drawn from the St. Mary River.

Figure 8.1 St. Mary River / Milk Watershed boundary.

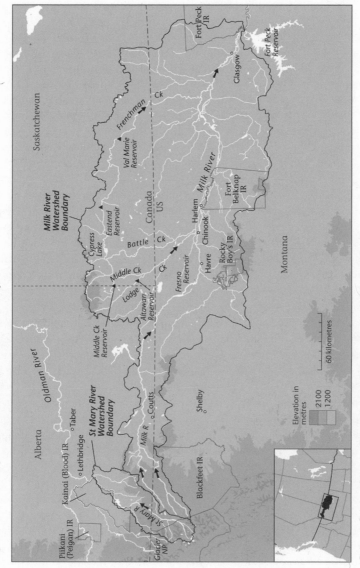

Source: Original Map by Eric Leinberger, University of British Columbia.

Figure 8.2 Lower St. Mary / Milk Watershed boundary and surrounding area.

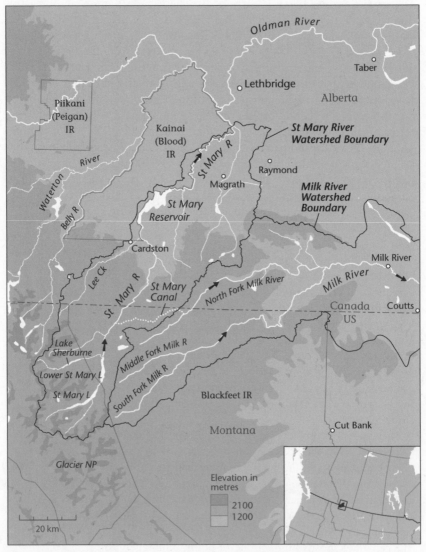

Source: Original Map by Eric Leinberger, University of British Columbia.

History

Mormon settlers first began using the St. Mary River as an irrigation source in Alberta in the late 1880s and early 1890s. The Alberta Railway and Irrigation Company (ARIC), with large landholdings based on railway land grants, acquired significant water rights on the St. Mary in 1897. Integrated development of the Waterton, Belly, and St. Mary Rivers occurred from the 1940s through the 1960s, with significant storage facilities at Waterton Dam and the St. Mary Reservoir. An interlinking canal system permitted the irrigation of 565,809 acres within eight irrigation districts, plus the 25,000-acre Blood Tribe Irrigation Project and some additional private irrigation. These facilities also provide recreational opportunities and water for municipalities. The region supplies some of the most productive farming in Alberta and is especially important for high-value specialty crops such as potatoes, sweet corn, and sugar beets. Irrigation provides stability and diversification to the region, and is of significant economic importance for the region and communities such as Taber, Magrath, Cardston, Raymond, and Lethbridge (Alberta Agriculture Food and Rural Development 2004).

In the United States, settlers constructed small irrigation projects and small dams along the Milk River in the 1880s and 1890s. However, the unreliability of the Milk River's water supply threatened the area's stability. The potential for irrigation[3] in the region had been recognized by US Army officers in the early 1870s in the Lower Milk River Valley. By 1891, an engineering report from the US Department of Agriculture indicated that it seemed feasible to dam the St. Mary Lakes at their outlet and build a canal to divert St. Mary River water into the North Fork of the Milk River before it crossed the border into Canada. These waters might then be conveyed by the channel of the Milk River through Canada and put to beneficial use in Montana's lower Milk River Valley. A further survey in 1901 examined the feasibility of a canal route entirely within the United States.

These possible projects led to concerns on the Canadian side of the border (especially given the potential for major irrigation developments and the existing water rights of the ARIC) that available water would be siphoned away in the United States before reaching Canada. In response, the ARIC built the aptly named "spite canal" (or ditch) in 1904 to show that it would be possible to convey water from the Milk River in Canada into the St. Mary Basin, essentially siphoning diverted St. Mary River water back into Canada further downstream. The canal was mostly built, but never put into full operation.

The St. Mary facilities are key to large-scale irrigation in the Milk River Basin in the United States. The system provides irrigation water for more than 110,000 acres of land, which in turn generate a significant portion of the cattle, hay, and alfalfa production in Montana; the water is regionally important for communities such as Havre, Chinook, and Harlem. As in Canada, the facilities also provide municipal water supply, as well as recreational opportunities for fisheries and water for wildlife (Azevedo 2003). The St. Mary Canal is frequently called the backbone or lifeline of the Montana "hi-line," a term used for northern Montana. For example, during the drought year of 2001, 95 per cent of the available water in the Milk River came from the St. Mary River (Azevedo 2003, 5).

Article VI of the Boundary Waters Treaty

The issues surrounding the development of the St. Mary were considered important enough to merit special treatment in the negotiation of the 1909 Boundary Waters Treaty between the United States and Great Britain, on behalf of the Dominion of Canada, resulting in what is now Article VI. Among other things, the BWT provided binational approval of the proposed St. Mary Canal and basin transfer, as well as establishing relevant water-apportionment rules. The text provides as follows:

> ... the St. Mary and Milk Rivers and their tributaries ... are to be treated as one stream for the purposes of irrigation and power, and the waters thereof shall be apportioned equally between the two countries, but in making such equal apportionment more than half may be taken from one river and less than half from the other by either country so as to afford a more beneficial use to each. It is further agreed that in the division of such waters during the irrigation season, between the 1st of April and 31st of October, inclusive, annually, the United States is entitled to a prior appropriation of 500 cubic feet per second of the waters of the Milk River, or so much of such amount as constitutes three-fourths of its natural flow, and that Canada is entitled to a prior appropriation of 500 cubic feet per second of the flow of St. Mary River, or so much of such amount as constitutes three-fourths of its natural flow.
>
> The channel of the Milk River in Canada may be used at the convenience of the United States for the conveyance of waters, while passing through Canadian territory, diverted from the St. Mary River.
>
> The measurement and apportionment of the water to be used by each country shall from time to time be made jointly by the properly constituted reclamation officers of the United States and the properly constituted irrigation officers of His Majesty under the direction of the International Joint Commission.

This is a complex provision with four key points. First, it pools the waters of the two rivers for the purposes of irrigation and power. Second, it effects an equal apportionment of the pooled waters; this apportionment is subject to a prior appropriation to Canada on the St. Mary River and to the United States on the Milk River. Third, it approves of the United States using the Milk River in Canada to convey diverted waters. Finally, it creates the administrative or supervisory jurisdiction of the IJC with respect to measurement and apportionment.

It is hard to overstate the uniqueness of this provision in the BWT. In general, the treaty established principles and procedures for settling future disputes between the United States and Canada. In only two cases did it effect an agreement for specific watercourses.[4]

Implementing Article VI: Construction of the Canal and the IJC's 1921 Order

The US Bureau of Reclamation completed construction of the St. Mary Canal in 1917. The canal was designed to carry a maximum flow of 850 cubic feet per second from the St. Mary River to the Milk River (International St. Mary – Milk Rivers Administrative Measures Task Force 2006). Lake Sherburne Dam, a primary storage feature of the project located in the headwaters of the St. Mary River, was operated for the first time in 1919, storing and then releasing 28,800 AF of water for project lands (Simonds 1999).

While Article VI effected an apportionment, the terms of that apportionment were hardly self-implementing. The IJC appointed two commissioners to look into the matter and report, and commission hearings followed between 1915 and 1921. Noting that the "Reclamation and Irrigation Officers have been unable to agree" about the implementation of Article VI (International Joint Commission 1921), the IJC issued an order in 1921. This order, which the commission adopted unanimously, is still in force today. The order specifies that:

> ... until this order is varied, modified, or withdrawn by the Commission, [the respective Officers shall] make jointly the measurement and apportionment of the water ... in accordance with the following rules:
> St. Mary River.
> I. (a) During the irrigation season when the natural flow of the St. Mary River at the point where it crosses the international boundary is six hundred and sixty-six (666) cubic feet per second or less Canada shall be entitled to three-fourths and the United States to one-fourth of such flow.

(b) During the irrigation season when the natural flow of the St. Mary River at the point where it crosses the international boundary is more than six hundred and sixty-six (666) cubic feet per second Canada shall be entitled to a prior appropriation of five hundred (500) cubic feet per second, and the excess over six hundred and sixty-six (666) cubic feet per second shall be divided equally.

(c) During the non-irrigation season the natural flow of the St. Mary River at the point where it crosses the international boundary shall be divided equally between the two countries.

Milk River.

II.   [This section of the order provides a parallel reciprocal provision giving the US preferential rights on the Milk River.]

III. The natural flow of the eastern (otherwise known as the Saskatchewan or northern) tributaries of the Milk River at the points where they cross the international boundary shall be divided equally between the two countries.

IX. In the event of any disagreement the said ... Officers shall report to the Commission, setting forth fully the points of difference and the facts relating thereto.

The order therefore contains parallel provisions for the two main rivers, notwithstanding their vastly different characteristics, both in total flow and the reliability of that flow. It is important to understand how the apportionment has worked in practice, and the 2006 report of the International St. Mary – Milk Rivers Administrative Measures Task Force provides a useful summative statement:

Over the period from 1950 to 2004 ... the US entitlement of the St. Mary River has averaged approximately 41 percent of the total annual flow (269,600 acre-feet or 332 600 dam3). Canada's entitlement of the St. Mary River has averaged approximately 59 percent of the total annual flow (388,000 acre-feet or 478 600 dam3).

The percentages tend to be opposite for the Milk River .... The apportionment for the Eastern Tributaries provides for an equal entitlement of the total annual flow (50 per cent) to each country. This amounts, on average, to approximately 57,600 acre-feet or 71 000 dam3 to each country.

Based on the past 55 years of record, application of the 1921 Order does not provide for equal entitlements to both countries of the annual flows of the St. Mary and Milk rivers. The combined entitlement for the St. Mary River,

Milk River and Eastern Tributaries results in approximately 45 percent going to the US and 55 percent going to Canada. (18–19, emphasis added)

Interests on either side of the border have radically different views about
whether this difference in actual water receipts is consistent with the terms of
the treaty. A basic view in the United States is that because Article VI treats the
two rivers as one combined water body and those combined waters must be
apportioned equally between the two countries, the United States is entitled
to half the combined water. Subsequent language specifies how that equal apportionment is to be carried out, and the order may provide further detail, but
the underlying principle of equal apportionment is paramount. Simply put, the
United States should receive half of the total water supply.

By contrast, the Canadian position, principally articulated by Alberta, is that
Article VI gives the United States the opportunity to take its share of the waters,
subject to the priority provisions. However, if the United States fails to do so,
perhaps because of inadequate canal or storage capacity, then the natural flow
regime will take over and the waters of the St. Mary River will continue to flow
down to Canada. This loss is not because of anything that Canada has done,
but because the United States has not taken full advantage of the opportunity
the BWT created. This, according to Canada, is not a breach of the treaty or the
terms of the order, but just the consequence of gravity. Canada and Alberta also
take the position that if the United States at any time passes more water down
to Canada than the terms of the order require, whether due to high flows or
inadequate infrastructure, Canada is not required to give credit for that surplus
delivery thus allowing the United States to take more than its apportioned share
later in the season.

The United States asked for reconsideration of the IJC Order of 1921 as early
as 1927, but the Commission split along national lines and the order remained
in force. This was highly unusual. The Commission's practice is to operate by
a consensus decision of its six commissioners, three from each country. In the
nearly 100 matters settled by the IJC since its inception in 1911, the commission has divided formally along national lines on only two occasions, both of
which involved the arid western region of North America (Blaney 2006, 303).

Administration of the 1921 Order

The administration of the 1921 Order is supervised by the accredited officers
of the two national governments. Over time, the officers have had to make
decisions on a variety of matters, including: when it is necessary to implement
apportionment rules on the various other streams the order covers (i.e., the

eastern tributaries), how to calculate natural flow, what should be the balancing period for reconciling obligations, and what are the implications of overdeliveries and underdeliveries. The last issue is perhaps the most significant, and the procedures that have evolved are as follows: (1) parties must balance deliveries twice a month, (2) a party (the upstream state) must compensate for any shortfall in the next delivery period (subject to the qualification discussed below), and (3) there is no credit to the upstream state for overdelivery.

The accredited officers first started systematically documenting these matters in 1975 in a document now known as the Procedures Manual for the Division of Waters of the St. Mary and Milk Rivers. The accredited officers have also agreed on some procedures that provide greater flexibility in applying the terms of the 1921 Order. The most significant of these is the letter of intent, which was originally agreed upon in 1991 and revised in 2001 (Accredited Officers of the United States and Canada 2001, 79). The 2001 version of the letter of intent begins with the twin recitals that Canada would prefer to use more than its share of Milk River water between June and September, and that the United States would prefer to use more than its share of the St. Mary River between March and May. The operative text then goes on to allow the United States to accumulate a deficit, or underdelivery, on the St. Mary River of up to 4,000 cubic feet per second – days (cfs-days) between March 1 and May 31. Canada can accumulate a deficit of 2,000 cfs-days on the Milk River between June 1 and September 15. The deficits may be offset, but must be equalized by October 31 each year. The net effect of the variation is to allow the United States to make better use of available capacity in the canal early in the season and to allow Canada to use some Milk River waters, including diverted St. Mary River waters, later in the season when the Milk River would have a very low or no natural flow. There has also been some flexibility in the administration of the arrangements for the eastern tributaries. In high flow years, mutual benefits are obtained by having Canada use available storage to underdeliver in the early part of the irrigation season, with the promise of balancing later in the season when there is more water demand downstream in the United States for irrigation purposes.

The Current Review

The Milk and St. Mary file was quiet for several decades until 2003, when Governor Judy Martz of Montana wrote to the chair of the US Section of the IJC requesting that the IJC review its 1921 Order. She stated that despite the Order having been in effect for 82 years, no one had yet evaluated it to determine if it was meeting the intent of Article VI of the Treaty. In subsequent correspondence, the governor elaborated three reasons for reopening the Order (Martz

2004). First, argued the governor, the 1921 Order failed to respect the requirements of Article VI(1) by treating the waters of the two rivers separately rather than as one, and it did not provide for equal apportionment of those waters, with Montana being particularly disadvantaged in drier years. Second, Governor Martz argued that circumstances had changed, including: (1) irrigation in the United States occurring in less than 60 per cent of the irrigated area contemplated in 1921, (2) increased Milk River use by Canada, (3) failure to apportion Rolph and Lee Creeks, and (4) that the full extent of Indian reserved rights was not known in 1921. Third, the governor pointed out problems with the apportionment procedure, specifically underestimation of Canada's uses of Milk River water and the absence of credit for surplus flows.

In response, Alberta fully supported the existing apportionment. The provincial government emphasised that investments had been made on the basis of the 1921 Order and that any changes to the sharing arrangement would have a devastating impact on the economy (Jonson 2003; Taylor 2004). Any "hold back" by Montana would create supply deficits (Alberta Agriculture Food and Rural Development 2004, 16). In response to Montana's arguments about the absence of credit for surplus flows, Alberta recognized that the United States was entitled to divert these waters and that the United States could undertake the necessary activities and investments without opening up the order. The province of Saskatchewan similarly emphasised the importance of certainty for the sustainability of investments (Forbes 2003), as did irrigation interests in both provinces (Hill 2003; Csabay et al. 2004). Alluding to the flexible arrangements on the eastern tributaries, Saskatchewan acknowledged (Forbes 2003) that the existing system allows water managers to develop mutually beneficial arrangements (Forbes 2003; Levy 2004).

In its detailed submission, the province (Alberta Environment 2004) responded to each of the state's three main arguments. There was, said Alberta, nothing new in Montana's claim that the 1921 Order failed to reflect the Treaty's language. All possible interpretations of Article VI(1) had been canvassed in the IJC's hearings between 1915 and 1921.[5] The Treaty drafters, as well as the relevant officials representing the two parties before the IJC, were perfectly aware that the St. Mary River offered a far more dependable flow than did the Milk River. Alberta also didn't accept the argument that there had been a material change in circumstances, stating instead that the parties, and in particular Alberta, had invested in the infrastructure necessary to make best use of its share of the waters based on the security the Order offered. In this case, Alberta emphasised the fact that each country received "excess flows," or flows beyond which it was entitled, "primarily due to the failure of each country to capture and utilize excess flow during snowmelt runoff or flooding events" (Alberta

Environment 2004, 7). In response to asserted problems with the administrative procedures, Alberta suggested that the parties had shown that they were committed to working through these difficulties, so this was no reason to open the Order for review. Finally, Alberta stated that the changed circumstances argument with respect to new projects (such as a possible storage project on the Milk River in Canada) was not convincing, since each party to the treaty was free to take appropriate steps to make full beneficial use of the water allocated to it.

For its part, the US government was considering rehabilitating the nearly century-old St. Mary Diversion and Conveyance Works, and developing emergency response plans in case of catastrophic failure prior to rehabilitation. The United States was also seeking settlements to resolve all outstanding water claims, including those by the Blackfeet Tribe concerning projects within the boundaries of the reservation (Ryan 2006). The capacity of the St. Mary Canal was diminished from its original design capacity of 850 cfs to approximately 670 cfs (Azevedo 2003). The US Fish and Wildlife Service noted that it would need to be consulted in any federal project to upgrade or enhance the existing project, to avoid causing jeopardy to bull trout, which is listed under the US Federal Endangered Species Act (Wilson 2004).

Two processes have considered these various concerns. The first was led by the IJC, and the second was initiated by the governor of Montana and the premier of Alberta. The two processes illustrate the various opportunities for problem solving available to different levels of government, but also emphasize the interlinkages between those levels of government. Native Americans participated in both processes, but were more directly involved on the US side than the Canadian side, where participation has largely been limited to providing written submissions.

Administrative Measures Task Force

The IJC held public consultations in June 2004 and met with tribal representatives, ultimately receiving more than 100 submissions through August 2004. Later in 2004, the IJC established an administrative-measures task force, directing it to examine and report on existing administrative procedures "to ensure more beneficial use and optimal receipt by each country of its apportioned waters." The task force was composed of four federal members, including the field representative to each government's accredited officer, two provincial and one state member, and a member suggested by US tribal representatives. The task force report (International St. Mary – Milk Rivers Administrative Measures Task Force 2006) identified computational improvements and flagged matters that may warrant attention in the future. It also identified potential benefits that could accrue to both countries if they implemented longer balancing periods and developed procedures that would allow for crediting surplus flows. Commentators on the US side of the border generally welcomed the report, particularly the consideration that it gave to lengthening the balancing period.[6] Canadian commentators were more cautious, but generally expressed willingness to continue exploring mutually beneficial options.[7]

With these promising results in hand, the commissioners met with the governor of Montana and the premier of Alberta. They followed up with a letter (Olson and Blaney 2007), in which the commission encouraged the governor and premier to meet and establish a small group "to explore the fundamental and interrelated issues of collaboration on the use and management of transboundary waters, cooperation on the rehabilitation of the St. Mary Canal and future arrangements for increasing the ability of each country to better access the full amount of water available to it under the current apportionment."

The Water Management Initiative (WMI)

The governments of Montana and Alberta did agree to establish a water management initiative (WMI) to explore opportunities for each jurisdiction "to better access the shared waters of the St. Mary and Milk River systems" (Alberta and Montana 2008). Launched in January 2009, the terms of reference for the WMI establish a 12-member Joint Initiative Team (JIT) of local water users and government officials from both sides of the border.[8] The JIT includes members from the state and provincial governments, the Blackfeet Tribe and the Fort Belknap Indian Community as part of the Montana delegation, and various watershed councils and organizations such as the Oldman and Milk River Watershed Councils on the Alberta side. The terms of reference (Alberta and Montana 2008) envisioned that the JIT would focus on access to water for irrigation and in-stream flow needs, as well as projects that could be jointly developed for the benefit of both parties, "specifically, rehabilitation of the St. Mary Canal." The JIT was allowed to recommend modifications to the letter of intent, the administrative procedures, and the Order of 1921 "if those instruments present a barrier to implementing preferred options" (Alberta and Montana 2008, 2).

The JIT was expected to conclude its business in 2010, but has yet to do so. The minutes of the meetings reveal that the discussions have been far-reaching, focusing on how to provide Montana with better access to its share of St. Mary River water and Alberta with better access to Milk River water. The JIT has met in formal session on at least 14 occasions and has considered more than 100 water management options (Montana – Alberta Initiative 2010b). One of the most important issues discussed is the concept of crediting for surplus deliveries, along with the related question of the appropriate balancing period.

Two particular proposals are described in the minutes of the February 2010 meeting (Montana – Alberta Initiative 2010a). Montana's version of the annual balance/credit based system is that the United States should be able to draw on and accumulate a single-fill balance of 32,000 AF on the St. Mary River. Meanwhile, Canada would be able to accumulate and draw upon a credit of 16,000 AF on the Milk River, provided that it does not draw more than 4,000 AF of St. Mary diverted flow rather than natural flow, with some reduction of this entitlement in dry years. The proposal also acknowledged that in drawing upon its credit, the United States would be obliged to meet certain in-stream flow requirements in the St. Mary River between the international boundary and the St. Mary Reservoir (Montana – Alberta Initiative 2010a, 3–4).

Alberta's proposal on the St. Mary River was similar, but capped the credit at 30,000 AF and required the United States to reduce its credit by 10,000 AF by June 1, or lose it. On the Milk River, Alberta proposed a credit cap of 15,000 AF, but also proposed a series of measures that would make it easier for the province to make use of that credit. Specifically, Alberta proposed that the actual size of the credit, subject to the cap, should be linked to the size of Montana's credit (surplus delivery) on the St. Mary River. Alberta would be able to take the lesser of its accumulated surplus, or 50 per cent of the US St. Mary River credit up to a maximum of 10,000 AF, but should be entitled to a minimum of 4,000 AF of US St. Mary River water in any year, regardless of credit accumulated (Montana – Alberta Initiative 2010a, 4–5).

The Alberta and Montana groups have discussed these proposals with the irrigation community, but at the time of this writing, it is not clear whether any compromise has emerged. Following an August 2011 meeting, both jurisdictions agreed to charge a technical team with conducting a detailed analysis of five previously reviewed options to determine the risks and benefits to each jurisdiction for consideration at the next meeting (Montana – Alberta Initiative 2011).

In-stream flows have not proven to be major issues in these negotiations. Montana is prepared to commit to maintaining minimum flows on the St. Mary River as a part of any arrangement that gives credit for surplus delivery. There are some concerns in Alberta that modifying flows on the Milk River represents a threat to the western silvery minnow, which is listed under the Species at Risk Act (SARA). However, this would likely only become a serious issue if the United States were to propose expanding the capacity of the canal to 1,000 cfs (Milk River Fish Species at Risk Recovery Team 2007). Similarly, as noted above, any federal action related to the St. Mary diversion in the United States would trigger consultations with the Fish and Wildlife Service with respect to the listed bull trout.

Alberta's Perspective – Concluding Comments

Alberta has two principal interests moving forward. First, on the St. Mary River, the province wishes to ensure that it can take maximum advantage of its already-constructed storage capacity on the combined St. Mary, Belly, and Waterton system. Thus, while it recognizes the United States' right to take its full share of the St. Mary River, Alberta has been reluctant to allow the United States to effectively use Canadian storage for free by accumulating a surplus in downstream reservoirs and making virtual deliveries from that storage later in the season when it would curtail Alberta's entitlement (i.e., by under delivering and allowing Alberta to balance by taking water from the earlier surplus or overdelivery). The current proposals show that Alberta is moving to accommodate Montana's interests, but subject to constraints to provide for in-stream flow requirements and to preclude the United States from carrying its surplus, and the discretion to use that surplus, too far into the irrigation season. Alberta recognizes that an expanded St. Mary Canal would allow the United States to better utilize its share of St. Mary River water without needing to accumulate a surplus. More generally, Alberta has an interest in the stability of the apportionment arrangements.

On the Milk River, Alberta is principally interested in ensuring water-supply security to existing irrigation and municipal interests, including the towns of Milk River and Coutts. The province recognizes that the variable nature of the Milk River means that the best security will come from assured access to St. Mary River water, transferred to the Milk River through the canal. Since neither the Treaty nor the 1921 Order allows Canada to access this water because it is not part of the natural flow, Alberta seeks a reasonable accommodation in return for providing credit for surplus deliveries on the St. Mary River.

Montana's Perspective – Concluding Comments

Montana is striving to find a way to improve its access to St. Mary River water. It has both a physical infrastructure option, and an administrative option. The physical infrastructure option includes investing in the rehabilitation, and perhaps the expansion, of the canal and possibly other irrigation infrastructure to increase efficiencies and reduce water losses. The principal challenge this option presents is the lack of access to capital, public or private. The irrigators are unwilling or unable to make the investment, and the timing is not opportune in terms of seeking monies from federal or state governments. The administrative option, which entails developing measures that allow a longer balancing period or crediting for surplus deliveries, would impose no additional costs on Montana.

Reflections and Considerations for the Future

The apportionment of the Milk and St. Mary Rivers is the longest-standing apportionment along the shared border between Canada and the United States. Although agreed to in principle in 1909, this experience shows that when it comes to effecting an apportionment, the devil is in the details. In particular, it is important to pay attention to issues such as the balancing period and giving credit for surplus deliveries. The principles underlying any such arrangements need to be applicable to all of the transboundary watercourses within the St. Mary River and Milk River watersheds, including the eastern tributaries as well as the main stems.

Ensuring that the infrastructure in place can continue to effect the apportionment is also necessary. Current economic trends mean that money is scarcer than ever, with a multitude of regular and deferred maintenance needs competing for diminishing funds. If infrastructure costs are to be shared to achieve mutual benefits, ideally this should be settled at the outset and based on those shared benefits.

As mentioned in chapter 6, climate change is shrinking the glaciers and snowfields that provide reliable summer flows downstream, and is changing accustomed precipitation patterns. Not only can these changes directly impact the amount and timing of flows, but they also call into question the validity of predicting future water flows based on more stable, long-term gage records. While research continues to provide useful new information, the level of uncertainty about future impacts remains high.

Uncertainty also plays a role in efforts to protect threatened species and provide for environmental purposes. Sometimes perceived as a "new" use or interest from those specified in the Treaty, environmental considerations introduce additional complexity into difficult and detailed discussions, but are becoming more broadly accepted as a necessary and previously overlooked part of the overall picture.

The state-provincial Initiative holds promise for better accessing the shared waters of the St. Mary and Milk Rivers under the framework provided by the Canada-US Boundary Waters Treaty and the IJC's Order of 1921. If history is any guide, it appears that it will be difficult to reconsider the provisions of the Order of 1921. Nevertheless, the experience with both the letter of intent and the eastern tributaries suggests that opportunities to realize mutual benefits within the order's framework are available. These mutual benefits could most likely be appreciated through cooperation between those most directly affected. The WMI provides a useful vehicle for this cooperation, bringing the state and the province to the forefront while including others that would be directly affected by the initiative.

In the longer term, those most directly affected (the province, the state, tribes, and irrigators on both sides of the border) may continue to be the ones best situated to address transboundary issues and concerns. The WMI provides a model for state-provincial cooperation for anticipated mutual benefit. Avenues for involvement of Indigenous people are important, and the inclusion of Native Americans (on the US side) and local watershed councils in the process of solving these issues is a significant evolution from the primarily federal-to-federal interaction the Treaty envisioned. The Treaty and the 1921 Order continue to provide a framework for federal and IJC involvement, which may be required to deal with issues – both foreseen and unexpected – that cannot be resolved at the more local levels.

NOTES

1 The tributaries include Lodge Creek, Battle Creek, and Frenchman Creek. There are some storage facilities on the Canadian side of the border that permit some regulation of the flow for flood control and irrigation purposes. The most significant of these facilities include Altawan Reservoir on Lodge Creek, Val Marie Reservoir and Eastend Reservoir on the Frenchman River, and Middle Creek Reservoir on Middle Creek. Cypress Lake provides storage for both Battle Creek and the Frenchman River.

2 This section is based on The Utilization of Water in the Milk River Basin within the Blackfeet Indian Reservation, Montana, 1960. This is one of a series of inspection reports carried out jointly by US and Canada state and federal officials between 1960 and 1965 in an effort to identify the scale of diversions on the south fork of the Milk before it crosses into Canada.

3 This next section draws on the following main sources Dreisziger 1974; Dreisziger 1975; Mitchner 1971; Purcell 1956. N.A.F. Dreisziger, "The International Joint Commission of the United States and Canada, 1895–1920. A Study in Canadian – American Relations" (PhD thesis, University of Toronto, 1974), esp. at 9 et seq. (This is the most comprehensive discussion of the Milk\St. Mary issue and its position in the negotiation of the Boundary Waters Treaty.); Dreisziger, "The Canadian-American Frontier Revisited: The International Origins of Irrigation in Southern Alberta, 1885–1909" (1975), Papers of the Canadian Historical Society, 211–229; E. Alyn Mitchner, "William Pearce and Federal Government Activity in Western Canada 1882–1904" (PhD thesis, University of Alberta, 1971). Mitchner deals with the Milk\St. Mary issues at pages 249–58. P. R. Purcell, Historical 176 Nigel Bankes and Elizabeth Bourget *Summary of Erosion of*

the Milk River, Department of Energy Mines and Resources, Inland Waters Branch, 1956. Resources, Inland Waters Branch, 1956.

4 The only other watercourse addressed specifically by the Treaty is the Niagara River, in Article V. Subsequent apportionments have all occurred pursuant to a "reference" to the IJC under Article IX. Under this procedure, the two governments pose specific questions to the IJC for its recommendation, which the governments implement (or not) as they see fit, sometimes by way of a new treaty, as in the case of the Souris River. For the original reference on the Souris, see http://www.ijc.org/conseil_board/souris_river/en/souris_mandate_mandat.htm#directive. For the treaty, see Agreement between the Government of Canada and the Government of the United States of America for water supply and flood control in the Souris River Basin, CTS 1989/36.

5 This point was previously contested in the 1930s, when the United States argued that the IJC had gone outside the questions debated during the hearings leading to the 1921 Order (Testimony of Senator Walsh to the International Joint Commission, 10–11 April 1931, Washington, DC); the IJC's consideration of the United States' concerns ultimately divided along national lines and therefore left the 1921 Order in place.

6 Comments received by the IJC are posted on its website at http://www.ijc.org/conseil_board/st_mary_milk_rivers2/smmr2_pub.php?language=english. Particularly notable on the US side were comments submitted by the state (Governor Schweitzer to Jewell 30 June 2006) and from the Bureau of Reclamation (Ryan to Herrington and Jewell, 30 June 2006). One of the co-authors of this paper (Bankes) provided comments available at: http://www.ijc.org/rel/pdf/smmr2/Bankes.pdf.

7 See for example, Guy Boutillier, Minister of the Environment to Herrington and Jewell, 30 June 2006, with attachment and asking the IJC to confirm that any future discussions should be within the framework of the 1921 Order. See also St. Mary River Irrigation District to Herrington, 16 June 2006. Saskatchewan suggested that special arrangements should be put in place for the Eastern tributaries, perhaps including the creation of Eastern Tributaries International Board modeled on the International Souris River Board: Duncan to the IJC, 27 June 2006.

8 The composition of the JIT does not reflect the IJC's suggestion in its October 2007 letter to the effect that the Committee include the field representatives responsible for implementing the 1921 Order. Neither do the Terms of Reference reflect the IJC's suggestions that the group might be referred to as an interim watershed council.

REFERENCES

Accredited Officers of the United States and Canada. 2001. "Letter of Intent to Better Utilize the Waters of the St. Mary and Milk Rivers." 8 February. In International St. Mary – Milk Rivers Administrative Task Force, Report to the International Joint Commission, 2006: 79. http://www.ijc.org/rel/pdf/SMMRAM.pdf.

Alberta Agriculture Food and Rural Development. 2004. "Irrigation Development in Alberta, Water Use and Impact on Regional Development, St. Mary River and 'Southern Tributaries' Watersheds." International Joint Commission submission, August. http://www.ijc.org/rel/pdf/83_stmary-milk_letter.pdf.

Alberta Environment. 2004. "Alberta's Submission to the International Joint Commission Respecting a Review of the IJC's 1921 Order on the Measurement and Apportionment of the St. Mary and Milk Rivers." Pub. No. I/974, August. Edmonton. http://www.ijc.org/rel/pdf/88_stmary-milk_letter.pdf.

Alberta and Montana. 2008. "St. Mary and Milk Rivers Initiative Terms of Reference." November. http://environment.alberta.ca/documents/MT-AB_St-Mary_Milk-Rivers_WMI_ToR.pdf.

Alyn Mitchner. E. 1971. "William Pearce and Federal Government Activity in Western Canada 1882–1904." PhD Thesis, University of Alberta.

Azevedo, Paul. 2003. "The Need to Rehabilitate the St. Mary Facilities." Department of Natural Resources and Conservation. http://dnrc.mt.gov/StMary/pdfs/stmarybackground.pdf.

Blaney, Jack. 2006. "An Overview of the International Joint Commission." In *Proceedings of the Fifth Biennial Rosenberg Forum on International Water Policy*, September, Banff, Canada.

Csabay, J., M. Zeinstra, and D. Hill. 2004. Letter to the International Joint Commission. Alberta Irrigation Projects Association, 30 August. http://www.ijc.org/rel/pdf/101_stmary-milk_letter.pdf.

Dreisziger, N. A. F. 1974. The International Joint Commission of the United States and Canada, 1895–1920: A study in Canadian-American relations. PhD Thesis, University of Toronto.

Dreisziger, N. F. 1975. The Canadian-American Frontier Revisited: The International Origins of Irrigation in Southern Alberta, 1885–1909, *Papers of the Canadian Historical Society*, 211–229.

Forbes, David. 2003. Letter to the International Joint Commission. Saskatchewan Watershed Authority, 16 December. http://www.ijc.org/rel/pdf/smmr/19-20031216-ForbesIJC.pdf.

Hill, David. 2003. Letter to the International Joint Commission. Alberta Irrigation Projects Association, 28 August. http://www.ijc.org/rel/pdf/smmr/14-20030828-HillIJC.pdf.

International Joint Commission. 1921. *Order in the Matter of the Measurement and Apportionment of the Waters of the St. Mary and Milk Rivers and their Tributaries in the United States and Canada.* Ottawa, 4 October.

International St. Mary-Milk Rivers Administrative Measures Task Force. 2006. *Report to the International Joint Commission*, April. http://www.ijc.org/rel/pdf/SMMRAM.pdf.

Jonson, Halvar. 2003. Letter to the International Joint Commission. Alberta International and Intergovernmental Relations, 24 July. http://www.ijc.org/rel/pdf/smmr/12-20030724-JonsonIJC.pdf.

Levy, Bruce. 2004. Letter to the International Joint Commission. Submission by the Government of Canada, 8 October. http://www.ijc.org/rel/pdf/107_stmary-milk_letter.pdf.

Martz, Judy. 2004. Letter to the International Joint Commission. Office of the Governor of Montana, 6 January. http://www.ijc.org/rel/pdf/smmr/20-20040106-MartzIJC.pdf.

Milk River Fish Species at Risk Recovery Team. 2007. "Recovery Strategy for the Western Silvery Minnow (*Hybognathus argyritis*) in Canada." *Species at Risk Act,* Recovery Strategy Series, Fisheries and Oceans Canada, Ottawa. http://www.sararegistry.gc.ca/virtual_sara/files/plans/rs_western_silvery_minnow_0208_e.pdf.

Montana-Alberta Initiative. 2010a. "Montana-Alberta: St. Mary and Milk Rivers." *Draft Notes of Joint Initiative Team Meeting #12.* 23–24 February. http://www.dnrc.mt.gov/wrd/water_mgmt/planning_activities/montana-alberta/meetings/notes2010_feb23-24.pdf.

Montana-Alberta Initiative. 2010b. "St. Mary and Milk Rivers." *Joint Status Report #14,* 21 July. http://www.dnrc.mt.gov/wrd/water_mgmt/planning_activities/montana-alberta/reports/joint_status_report14.pdf.

Montana-Alberta Initiative, 2011. Highlights of August 16, 2011 Meeting. http://environment.alberta.ca/01257.html

Olson, A., and J. Blaney. 2007. Joint letter to Governor Brian Schweitzer and Premier Ed Stelmach. In Appendix 1 to Initiative Terms of Reference Terms of Reference, 19 October. http://environment.alberta.ca/documents/MT-AB_St-Mary_Milk-Rivers_WMI_ToR.pdf.

Purcell, P. R. 1956. *Historical Summary of Erosion of the Milk River, Department of Energy Mines and Resources, Inland Waters Branch.*

Ryan, Mike. 2006. Statement of Mike Ryan, Regional Director – Great Plains Region Bureau of Reclamation, U.S. Department of the Interior, before the Energy and Natural Resources Committee, U.S. Senate, on *S. 3563 The St. Mary Diversion and Conveyance Works and Milk River Project Act,* 1 September. http://www.usbr.gov/newsroom/testimony/detail.cfm?RecordID=761.

Simonds, W.J. 1999. "Milk River Project, Research on Historic Reclamation Projects." U.S. Bureau of Reclamation History Program, Denver, Colorado. http://www.usbr.gov/projects/Project.jsp?proj_Name=Milk%20River%20Project&pageType=ProjectHistoryPage#Group153095.

Taylor, Lorne. 2004. Letter to the International Joint Commission. Alberta Environment, Office of the Minister, 13 July. http://www.ijc.org/rel/pdf/43_stmary-milk_letter.pdf.

US Bureau of Reclamation. 2004. *Regional Feasibility Report, North Central Montana.* Billings, Montana, October. http://www.usbr.gov/gp/mtao/ncmrfr.pdf.

Wilson, Mark. 2004. "Technical Comments Pertaining to Bull Trout and Critical Habitat for the St. Mary River DPS." Letter to the International Joint Commission, 26 July. http://www.ijc.org/rel/pdf/64_stmary-milk_letter.pdf.

9 Devils Lake and Red River Basin

NORMAN BRANDSON AND ROBERT HEARNE

The Devils Lake Basin is a 9,868-square-kilometer (3,810 square-mile) closed basin in northeast North Dakota that drains into the largest freshwater lake in the state (see Figure 9.1). In general, water enters the basin through precipitation and exits through evaporation. Water levels in Devils Lake rise and fall dramatically, leading to cyclical periods of flooding. At least twice in the last 4,000 years, water has spilled out of the basin and into the Sheyenne River, which flows into the Red River of the North, which in turn flows into Lake Winnipeg and the Hudson Bay. Due to its lack of natural outflow, the lake is characterized by high salinity and high levels of sulphates and suspended solids.

Tensions in the region have increased considerably since 1993 when a rapid increase in lake levels greatly expanded the surface area of Devils Lake, flooding hundreds of miles of farmland, damaging and destroying roads, and threatening numerous communities including Devils Lake, the area's largest city. To mitigate this flooding, which has become an urgent priority for politicians and water managers, the state of North Dakota has initiated interbasin transfer projects. North Dakota's unilateral action, which the province of Manitoba has strongly opposed, has led to a protracted transboundary water conflict – arguably the most contentious of those this volume discusses. The failure of these neighbours to activate the established Canada-US dispute-resolution process has led to increased rhetoric and mistrust among all involved parties. Manitoba stresses the potential ecological damages from the transfer of harmful aquatic biota into the Hudson Bay Basin; North Dakota has downplayed the dangers of biota transfer, focusing on the state's current losses from flooding and the potentially catastrophic overflow into the Sheyenne River. Meanwhile, despite mitigation efforts, lake levels continue to remain near historical highs and discourse continues to escalate as of 2012.

Figure 9.1 Map of Devils Lake and Red River Basin.

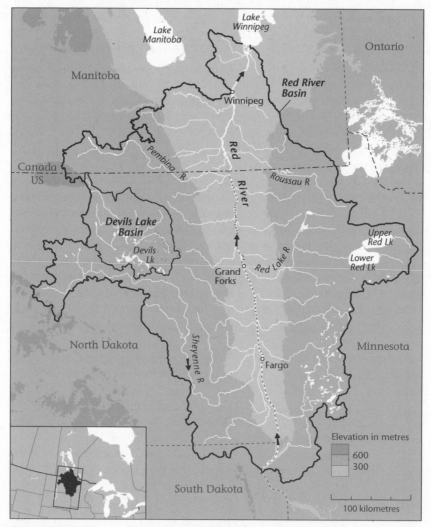

Source: Original Map by Eric Leinberger, University of British Columbia.

This chapter provides a concrete example of the much-discussed phenomenon of moving away from the International Joint Commission's (IJC) references. The latter part of this chapter addresses the causes and consequences of this particular matter not being referred to the IJC. We urge readers to bear this in mind when reflecting on the trend away from IJC-mediated disputes and when weighing the pros and cons of federal-to-federal level discussions.

Physical and Social Characteristics

The Red River Basin has a drainage area of nearly 48,000 square miles and a population of approximately 1.3 million, with 670,000 people living in the city of Winnipeg, Manitoba. Although the predominant land use is agriculture, the population is generally concentrated in the larger metropolitan areas on the Red River. The primary economic activity in the region is farming: wheat, barley, sunflowers, potatoes, corn, soybeans, and sugar beets are the principal crops (Hearne 2007; Krenz and Leitch 2003).

The Devils Lake Basin has a population of nearly 22,000, including more than 6,000 residents of the Spirit Lake Tribal Reservation on the southern shore of Devils Lake. The basin is adjacent to the centre of the North American continent and is thus characterized by short, warm summers and long, cold winters. The winters are relatively dry, with three-quarters of the average annual precipitation of 17 inches falling between April and September. However, this winter precipitation and subsequent snowmelt are key determinants of the spring water-level increase in Devils Lake. The increasing frequency and intensity of summer-storm events over the past two decades has proven to be the critical factor influencing the rise in water level. (See Figure 9.2.)

Figure 9.2 Prehistoric hydrology of Devils Lake.

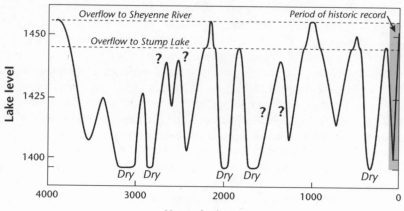

Source: North Dakota State Water Commission.

The cyclical pattern of water-level increase and decrease in the Devils Lake Basin precedes anthropogenic global climate change. However, researchers suggest that increasing greenhouse-gas levels and subsequent climate changes will generate additional contributing influences on Devils Lake water levels. In general, global climate change is expected to cause higher mean temperatures and more extreme climate events, such as droughts and floods. Higher mean temperatures in the Great Plains are expected to induce higher levels of evaporation, but increased precipitation is also predicted for this region (Karl et al. 2009). Thus, superimposed upon the cyclical Devils Lake weather patterns are the two competing influences of projected increased precipitation and evaporation.

Manitoba and North Dakota Positions

Although the state of North Dakota and the province of Manitoba share a similar semi-arid climate and a large portion of the Red River Valley, the differences are much greater than the similarities. Even before the current wet cycle – driven flooding of Devils Lake, each jurisdiction had its own perspective on water issues. North Dakota is relatively dry and has few lakes, and it shares its two major rivers, the Red and the Missouri, with several other jurisdictions. Manitoba, on the other hand, has styled itself the "Land of 100,000 Lakes" (an underestimate) and contains the downstream catchment of the entire 1.5 million-square kilometre Hudson Bay drainage basin, including the Assiniboine, Red, Saskatchewan, Churchill, and Winnipeg Rivers.

From its inception as a state in 1889, North Dakota has focused on moving water to irrigate and drain farmland, meeting municipal and industrial demand, and reducing flooding. Numerous proposals to gain federal funding for infrastructure projects have been developed as part of the Garrison Diversion project, the Devils Lake flooding problem, and flooding in the Red River Basin. Manitoba's preoccupation has been flood control. The province is downstream from four states and four provinces, and the quantity and quality of water crossing its borders is an ongoing concern. A series of major flood-control projects have been constructed and expanded over the past half-century, including the Shellmouth Dam on the Assiniboine River, which diverts a portion of Assiniboine flows into Lake Manitoba and the Winnipeg Floodway. In the far north, hydroelectric power generation is the predominant water use. In the mid-1970s, the province diverted a large portion of the Churchill River into the Nelson River to augment flow through hydroelectric plants, a project that remains controversial.

From the North Dakota perspective, the main issues to be addressed are the serious flooding, property loss, and damage that has already occurred and will continue to occur as long as Devils Lake continues to rise. The state of North Dakota has accepted responsibility to reduce the loss of productive farmland and valuable infrastructure. The NDSWC has stated that the risk of a natural overflow into the Sheyenne River, a risk that is between 8 and 20 per cent probability without any outlet, is high, and this overflow would be devastating (Fridgen 2011). Building dikes and passively hoping that the wet cycle will end before natural overflow occurs does not meet the needs of the state. Restoring drained wetlands cannot be expected to mitigate the flood damages sufficiently. Therefore, the only remaining option is to move water out of the basin via an artificial outlet. Even though flow and water quality constraints will limit the amount of water that can be released, and optimistic forecasts show mere inches being taken off the lake, depending on when the wet cycle ends, those inches could be critical.

North Dakota considers concerns about water quality and invasive species to be groundless, since they are based on what they consider "faulty science." North Dakota feels that a certain level of biota transfer across basin boundaries is inevitable and easily cites numerous natural overflows across the Devils Lake – Red River boundaries (Peterson Environmental Consulting 2002). From the state's perspective, the International Joint Commission is not considered helpful in this instance because of how long it has taken to resolve issues previously referred. Therefore, it is best for North Dakota not to consider a Devils Lake outlet to be a Boundary Waters Treaty (BWT) issue. The Spirit Lake Nation concurs with the North Dakota position that flooding should be alleviated (Pearson 2010).

From the Manitoba perspective, the release of water from Devils Lake poses a risk to provincial waters, particularly Lake Winnipeg. Devils Lake has been isolated from the Red River and Lake Winnipeg watersheds for almost two millennia. Fish species such as walleye, pike, and striped bass have been introduced to the lake after being reared in hatcheries in the Missouri watershed. Recreational fishers, many from outside the Red River Basin, bring bait fish to the lake and whatever else might be carried with their boats and bilge water. Manitoba accepts the scientific support of a nontrivial risk of invasive species transfer as credible and does not consider water quality concerns to be misguided since significant amounts of total dissolved solids (sulphates) and nutrients (phosphorous) will be released from Devils Lake. The Manitoba stance also suggests that no outlet options will solve the Devils Lake problem. The probability of a natural overflow is low, and should it happen, the result would be similar to the overflow between Devils and Stump lakes and should not be considered catastrophic. Numerous options could reduce or eliminate the risk of a natural overflow.

Pressures: Floods and Droughts

The most pressing water management concerns in the Red River Basin are the interrelated issues of drainage, flooding, and water quality. The basin is extremely flat, so excess precipitation and spring snowmelt often lead to periodic flooding. This flooding requires farmers and nearby cities to invest in drainage systems to allow for crop production and avoid extreme flood events, respectively. These extensive agricultural and urban drainage systems lead to nonpoint-source pollution, which traditionally has not been a major concern (Stoner et al. 1998). Recently, however, algal blooms in Lake Winnipeg have threatened the Earth's eleventh largest freshwater body, and much of this water quality deterioration is attributable to phosphorus and nitrogen coming from the US portion of the Red River (Gunderson 2010). Another feature of the water quality and geography of the Red River Basin is that its relative isolation has protected it from many invasive aquatic species, including Eurasian milfoil and the zebra mussel, which have invaded other freshwater ecosystems, such as the Mississippi River and the Great Lakes.

The basin is also vulnerable to extended drought. During dry periods, particularly the 1930s Dust Bowl, water flow in the Red River significantly declines. The basin's flat topography makes it impractical to store water in surface reservoirs, so cities such as Fargo-Moorhead that lie in the upper section of the mainstream Red River are vulnerable to severe water shortages during times of drought.

The challenges facing Red River water managers often imply proposed solutions that differentially impact upstream and downstream users. The Devils Lake outlet controversy certainly is just one of the issues that caused disagreement between North Dakota and Manitoba water management authorities. One long-standing disagreement involves proposed and actual Garrison Diversion projects to bring Missouri River water to the Red and Souris River Basins.

The Garrison Diversion efforts date back to the construction of a series of large, multipurpose dams and reservoirs along the Missouri River as part of the Pick-Sloan Missouri Basin Program. In addition to flood control, recreation, hydroelectric generation, and downstream navigation, local states were promised irrigation development to compensate them for the loss of valuable riparian land. Between 1946 and 1966, six reservoirs were constructed on the mainstream Missouri River. This included Lake Sakakawea, the third-largest constructed reservoir in the United States, located in northeast North Dakota. Originally, North Dakota was promised 1.275 million acres of irrigation development. However, studies demonstrated that soils in northwestern North Dakota did not meet federal irrigation standards. Thus, since 1957, plans and studies have focused on transferring water from the Missouri Basin to the Red River Basin in eastern North Dakota. Since 1965, these plans have included municipal and industrial water, recreation, and irrigation.

Mitigation as a Transboundary Irritant

Although several hundred million dollars had been expended on Garrison Diversion works, Canadian concerns about water quality and invasive species ultimately resulted in a joint reference to the IJC (for a detailed discussion of the IJC, see box 1.1 and chapter 4, this volume). In 1977, the IJC recommended against transferring water between the Missouri and Hudson Bay drainages, and stipulated that no such transfer occur unless both national governments agreed that it could be done safely. There were subsequent changes to the authorizing legislation for Garrison, culminating in the Dakota Water Resources Act of 2000, which renamed the project the Red River Valley Water Supply Project. This project and the similar Northwest Area Water Supply (NAWS) project, which proposes bringing potable Missouri River water to the Souris River Basin communities in north central North Dakota, continue to be pursued and promoted in part as "repayment" for "prime ND farmland" flooded by the Garrison and Oahe reservoirs.

As of 2012, both of these Missouri River interbasin transfers have federally mandated Environmental Impact Assessments (EIAs) under a stage of public comment. With federal funding, North Dakota has started constructing pipelines for the NAWS system. In a case filed by Manitoba, the US District Court found the NAWS EIA to be insufficient in dealing with the transfer of aquatic species. The United States Bureau of Reclamation initiated a supplemental EIA. In 2007, the Bureau of Reclamation and the Garrison Diversion Unit released a final EIA for a proposed project that would bring Missouri River water to the Red River Basin communities for potable water supply. This project is currently awaiting approval for federal funding and the use of Missouri River water. Given the interests of downstream states in Missouri River management, it is incorrect to assume that any North Dakota initiative to transfer large quantities of water away from the Missouri River Basin is politically feasible in Washington. For instance, the growing demand for water in North Dakota's oil-producing region near Lake Sakakawea may reduce the push to transfer water out of the Missouri basin – an issue at the water/energy nexus discussed in chapter 6. In December 2010, the US Army Corps of Engineers (USACE) released a draft plan to allocate surplus Lake Sakakawea water temporarily to the North Dakota petroleum-drilling industry during the 10-year period before they complete a comprehensive evaluation of Missouri River dam water and dam management (USACE 2010).

Another cross-border conflict that has coincided with the Devils Lake outlet controversy involves a 30-mile border road, which also serves as a dike and contains water from the periodically flooded Pembina River in North Dakota. This controversy dates back to the 1940s when the structure was built. As part of a 2005 Canada-US agreement, culverts were to be constructed to relieve flooding. During the period of disagreement over the Devils Lake outlet, North Dakota has often asserted that these culverts have been blocked. This road has been considered a mechanism for less-than-equal retaliation by Manitoba for the Devils Lake diversion and a reciprocal demonstration of the potential consequences of unilateral action and noncooperation.

Escalating Conflict: Chronology of Events since Recent Lake Level Rise

Disagreement over the Devils Lake outlet is a direct result of the latest shift in regional precipitation, dating back to 1980. The four-inch increase in summer and fall rainfall has caused a lake level rise of 31 feet (9.45 metres) since 1993 (Vecchia 2010) (see Figure 9.3). This rise in lake water levels has also caused the volume of water in the lake to increase by six times. The land area covered by the lake has consequently increased by more than 215 square miles (North Dakota State Water Commission 2010). In addition to the loss of valuable farmland, roads have needed to be moved and raised. In 1996, the Federal Emergency Management Agency (FEMA) initiated a buyout program to compensate landowners for lost property. In addition, a number of levees have been constructed to protect the city of Devils Lake, which has a population of 7,000. Currently, efforts are ongoing to protect the city up to a lake level of 1,460 feet (445 metres) above mean sea level (amsl), which is just above the level at which the lake will naturally spill into the Sheyenne River.

In 1997, USACE was authorized to study the feasibility and the environmental impacts of an outlet from Devils Lake to the Sheyenne River. Construction of this proposed outlet was subject to assurances from the secretary of state that the Canada-US Boundary Waters Treaty would not be violated and that the project was in compliance with existing water-quality standards. The USACE estimated potential damages of $900 million if the lake continued to rise to the natural overflow level of 1,458 feet. The USACE chose a 22-mile, 300-cubic-feet-per-second (cfs) outlet from Pelican Lake to the Sheyenne River as the preferred alternative. This option was considered the least environmentally damaging because water in the upstream Pelican Lake has lower levels of dissolved solids and sulphates than the lower stretches of the lake (Elstad 2002). The estimated cost of the plan was US$200 million.

Figure 9.3 Historical record of Devils Lake elevation as of May 2011.

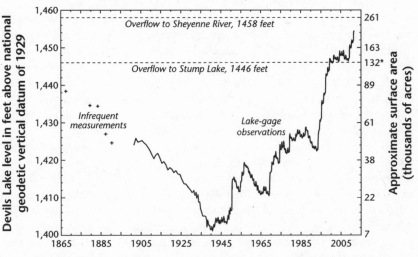

*Surface areas above elevation 1446 feet include Devils Lake and Stump Lake combined.

Source: USGS North Dakota Water Science Center.

In 2002, the US Department of State requested a joint referral of the Devils Lake flooding and outlet proposal to the International Joint Commission. Canada declined, stating that the Environmental Impact Assessment (EIA) was still under review and that a preferred alternative had yet to be chosen. Canada requested a broader referral that would consider other inter-basin transfers, including the proposed Garrison transfer (Kergan 2002).

The state of North Dakota chose not to wait for federal approval and funding for the outlet, and instead authorized the State Water Commission (NDSWC) to construct a 14-mile, 100-cfs, $28 million emergency outlet. Since federal funds were not used, an EIA of this project was not conducted. The outlet was constructed, despite opposition from Manitoba, Minnesota, Missouri, the Great Lakes States, a local interest group, the Friends of the Sheyenne. In compliance with state water quality protocols, the state outlet was initially given a permit from the North Dakota Department of Health (NDDOH) that restricted operation of the outlet to the following conditions: (1) only during the period of May through November; (2) only when lake levels are above 1,445 feet amsl; (3) only when downstream sulphate concentrations are less than 300 mg/l; and (4) only when total downstream suspended solids concentrations are less than

100 mg/l. Upon construction of the state outlet, Canada requested a referral to the IJC. The US government refused.

In August 2005, upon completion of the North Dakota state outlet project and upon imminent utilization, a joint press statement announced an agreement between the United States, Canada, North Dakota, Minnesota, and Manitoba. This was an unsigned signal of cooperation. The accord stipulated that North Dakota install a filter at the diversion before outlet operation to protect downstream areas from biota transfer. A more advanced filtration system was to be installed later. In addition, a rapid bioassessment would evaluate the risk of biological contamination from Devils Lake into the Red River.

After the released agreement, a sand-and-gravel filter was established at the outlet. A report from a multijurisdictional team was released in July 2005; the report found that in a limited time of sampling, no biota of concern were detected in Devils Lake. In September 2005, another report on fish pathogens in Devils Lake was released. No viral fish pathogens were detected, and overall fish health was good. Certain bacteria, pathogens, and parasites were encountered, but these were considered common to other North American waters (Hudson and Peters 2005).

Since 2005, the operation of the outlet has been severely restricted due to sulphate levels in the Sheyenne River. During 2006 and much of 2007, the upper stretches of the Sheyenne had high levels of naturally occurring sulphates. Because of this, the outlet only operated for 38 days during these years (Nicholson 2007). In August 2006, the NDDOH's permit was amended to allow for discharges when the Sheyenne was not frozen and had sulphate levels less than 450 mg/l. However, the amount of naturally occurring sulphates often exceeded this level.

As water levels and corresponding damages continued to increase in the Devils Lake Basin, the NDDOH authorized increasingly relaxed restrictions on operating the outlet. In June 2009, the NDDOH cited a US Environmental Protection Agency (EPA) ruling that transfers of unchanged water from one body to another did not require a permit and ruled that NDSWC did not need a permit to operate the outlet. The White House reaffirmed the lack of need for EPA oversight in these transfers in November 2010, which occurred without consultation with Canada and after the 2005 four-party talks had resumed in October 2010. In July 2009, the NDDOH ruled that the upper stretches of the Sheyenne River were not used for municipal drinking water, and therefore, could accept sulphate concentrations of up to 750 mg/l despite USEPA recommendations that drinking water should not have sulphate levels exceeding 250 mg/l. Simultaneously, the NDSWC began to expand the pumping capacity for the outlet to increase the flow rate to 350 cfs. In March 2011, North Dakota unilaterally announced plans to build an additional 250 cfs outlet from East Devils

Lake to the Tolna Coulee. The announcement did not include any mention of a filter to reduce the risk of biota transfer (North Dakota State Water Commission 2011). Canada opposed each of this measures designed to allow for greater flow through the outlet.

Conclusions

What is an outside observer to make of this debate? Surely residents in the Devils Lake Basin ought to pursue whatever options they can to alleviate a major tragedy. Yet why should Manitoba residents have to bear the risk of unintended consequences of diverting water out of this closed basin? It is increasingly apparent that neither jurisdiction is likely to accept the views of the other. Both jurisdictions have criticized unfavourable scientific studies and deemed certain low-risk events unlikely and/or unimportant. This is a classic upstream-downstream conflict. The BWT anticipated such conflicts over watersheds shared by the two countries and offers a conflict-resolution mechanism; but is the BWT, of course, is of little help if it is not used.

Although the position of each jurisdiction is eminently reasonable from its own perspective, the prospect of a negotiated settlement that satisfies all interests is unlikely and, at least outside the BWT, all leverage rests with the upstream jurisdiction. A combination of factors makes a mutually satisfying agreement unlikely: an assertion that Canada had already been offered an IJC reference on Devils Lake but "declined" (not entirely true, since Canada responded that at the time, there was no specific project to evaluate and a reference was perfectly appropriate when there was a specific proposal on the table); an inconsistent approach from the government of Canada, which ultimately agreed to circumvent the BWT process for a political negotiation moderated by the President's Council on Environmental Quality; and a refusal of the state department to formally respond to Canadian requests for a joint reference.

The value of an IJC reference would have been:

- An objective assessment, given the long history of references for which the two governments have only twice failed to agree to IJC recommendations.
- A scientific review of the evidence of environmental risks under alternative policies.
- An examination of a wider range of options, such as water treatment, pumping to the James River system, a controlled Tolna Coulee outflow if natural overflow occurs, and upper-basin storage through wetland restoration.

- Potentially timely recommendations, given that the 1997 Red River Valley Flood reference produced a preliminary report in less than a year.
- Increased "elbow room" for either or both jurisdictions to be able to accept actions that might not perfectly align with local politics.
- An improved climate for discussion of other Manitoba – North Dakota water issues.

The damage resulting from avoiding a reference has been:

- A dangerous precedent set whereby states and provinces, by exerting political pressure, can avoid the BWT process.
- Increased rhetoric from both sides of the border and a disrespect for science-based risk assessment.
- An untested assertion that the IJC process represents an unacceptable delay in decision-making.
- A further precedent establishing a political negotiation process that produced an "agreement" that has been nonbinding and with commitments that have not been respected.

Other instances in the recent past show how a state or province has thwarted the BWT by exerting pressure on its national government. Although the Canadian provinces and US states are the proximate cause of this recent ascendance of parochial interests, surely both national governments have to accept responsibility for the continued conflict that occurs when mediation fails. The BWT is an express and far-sighted recognition that purely local interests cannot be the only criteria to adjudicate transboundary water issues between two countries. In the case of Devils Lake, the failure to allow for mediation has created a legacy of bitterness between two neighbours, which will inevitably taint future water relations and provide an unfortunate precedent to other governments contemplating water projects that may pose a risk across the border.

Some cooperation between North Dakota and Manitoba does continue. The International Red River Board – one of the IJC's IWI pilot projects (see chapter 4) – serves as an autonomous coordinating institution, continues its efforts to foster cooperation, and seeks agreement among riparian constituencies and decision makers. Certain efforts are being made to reduce US contribution to the nitrogen and phosphorous levels in the Red River, which generate the algal blooms in Lake Winnipeg. Overall, the spirit of compromise that is required for the effective management of shared water has been imperiled by the discord over Devils Lake. The mutual benefit that comes from well-managed shared waters implies that the two constituencies should discard much of the negative

rhetoric of the Devils Lake discord and find common ground through cooperation on issues of agreement.

REFERENCES

Elstad, S. 2002. "Chemical, Physical and Biological Characterization of Devils Lake
1995–2001." Report by North Dakota Department of Health Division of Water
Quality. http://www.swc.state.nd.us/4dlink9/4dcgi/GetSubCategory Record/
Devils%20Lake%20Flooding/Studies%20and%20Reports. Accessed February 2011.

Fridgen, P. 2011. "More Help on the Way for Devils Lake." *The Oxbow*, January. http://
swc.state.nd.us/4dlink9/4dcgi/GetContentPDF/PB-1899/ OxbowJan2011.pdf. Accessed March 2011.

Gunderson, D. 2010. "Red River Pollution Threatening Lake Winnipeg." *MPR News*,
17 June. http://minnesota.publicradio.org/display/web/2010/06/17/ lake-winnipeg/.
Accessed March 2011.

Hearne, R.R. Dec 2007. "Evolving Institutions in the Red River Basin." *Environmental
Management* 40 (6): 842–52. http://dx.doi.org/10.1007/s00267-007-9026-x.
Medline:17912585

Hudson, C., and K. Peters. 2005. "Survey of Specific Fish Pathogens in Free-Ranging
Fish from Devils Lake, North Dakota." Bozeman Fish Health Center Technical Report 05–02. http://www.swc.state.nd.us/4dlink9/4dcgi/GetSubContent PDF/PB-835/
fish2005.pdf. Accessed February 2011.

Karl, T.R., J.M. Melillo, and T.C. Peterson, eds. 2009. *Global Climate Change Impacts
in the United States*. Cambridge: Cambridge University Press. http://www.
globalchange.gov/ publications/reports/scientific-assessments/us-impacts/
download-the-report. Accessed February 2011.

Kergan, M. 2002. Letter to Ambassador Marc Grossman. http://www. swc.state.
nd.us/4dlink9/4dcgi/GetSubContentPDF/PB-86/IJC_Referral_MB_Response .pdf.
Accessed February 2011.

Krenz, G., and J. Leitch. 2003. *A River Runs North: Managing an International River*.
Red River Water Resource Council.

Nicholson. 2007. "Devils Lake Outlet Shut Down for Winter." The Bismarck Tribune,
28 October. http://www.bismarcktribune.com/news/state-and-regional/ article_
e8cb0a25- b237–584b-b14b-3a0191e7ed6b.html. Accessed February 2011.

North Dakota State Water Commission. 2010. "Devils Lake Flood Facts." http://www.
swc.state.nd.us/4dlink9/4dcgi/GetSubCategoryRecord/Devils%20Lake %20Flood-
ing/Outlet. Accessed February 2011.

North Dakota State Water Commission. 2011. "State Moving Forward on East
Devils Lake Outlet." *The Oxbow*, April. http://swc.state.nd.us/4dlink9/ 4dcgi/
GetContentPDF/PB-1946/OxbowApr2011.pdf. Accessed March 2011.

Pearson, M. 2010. Testimony at the U.S. Senate Budget Committee Field Hearing held in Devils Lake, ND on 8 July. http://budget.senate.gov/democratic/ index.cfm/ committeehearings?ContentRecord_id=51a49371–4cd6–4738-b8bc-94fc373c8 8bf&ContentType_id=14f995b9-dfa5–407a-9d35–56cc7152a7ed&Group_ id=d68d31c2–2e75–49fb-a03a-be915cb4550b. Accessed June 2011.

Peterson Environmental Consulting. 2002. "Biota Transfer Study: Devils Lake Flood Damage Reduction Alternatives." Report prepared for St. Paul District United States Army Corps of Engineers. http://www.swc.state.nd.us/4dlink9/4dcgi/Get Content-PDF/PB-504/EIS_BIOTA_FINAL_REPORT.pdf. Accessed March 2011.

Stoner, J.D., D.L. Lorenz, R.M. Goldstein, M.E. Brigham, and T.K. Cowdery. 1998. "Water Quality in the Red River of the North Basin, Minnesota, North Dakota, and South Dakota, 1992–95." US Geological Survey Circular 1169. Reston, VA: US Geological Survey, Water Resources Division.

USACE. 2010. "Garrison Dam/Lake Sakakawea Project North Dakota Draft." Surplus Water Report. http://www.nwo.usace.army.mil/html/pd-p/Sakakawea _SWR_Public_Draft.pdf. Accessed January 2011.

Vecchia, S. 2010. "Future Flood Risk for Devils Lake." USGS presentation to the Devils Lake Flood Summit, 3 May. http://www.swc.state.nd.us/4dlink9/ 4dcgi/GetSubCategoryRecord/Devils%20Lake%20Flooding/Outlet. Accessed February 2011.

10 The Flathead River Basin

HARVEY LOCKE AND MATTHEW MCKINNEY

This book explores the past and future prospects for transboundary water governance between Canada and the United States. It also seeks to address the key challenges and opportunities facing water managers, politicians, and water users. This chapter illustrates that the challenge of water governance can go beyond water itself and include the entire watershed – water and land. In this case, efforts to resolve issues in the transboundary Flathead River Basin have not only triggered but also gone beyond the water governance framework contemplated by the Boundary Waters Treaty (BWT) and the International Joint Commission (IJC). (For a detailed discussion of the IJC, see box 1.1 and chapter 4.)

A dispute over a proposed Cabin Creek coal mine in the British Columbia portion of the Flathead led the two countries to seek an IJC reference, which resulted in report in the 1980s (IJC 1988). But time revealed that intervention was not sufficient to resolve the issue of resource development in an upstream riparian country. The conflicts flared up again in the first decade of the twenty-first century when another mine was proposed for a different tributary of the river, further upstream in British Columbia. This time, the conflict remained at the political level and the IJC was not invited to become involved. The dispute recurred because conflict over the Flathead River essentially relates to land use and differing goals for the development or protection of natural resources in the watershed. Efforts to reconcile those conflicting land-use regimes have been the continuing focus of those who seek to resolve this (nominal) water dispute. This chapter therefore takes a slightly different approach from the other chapters: it looks not only at water, but also at the interconnection between water governance and broader land-use questions in the Flathead watershed.

The Conflict over the Flathead River

The conflict over the Flathead River, which flows from British Columbia (BC), Canada, to Montana, USA, began in the 1980s and became acute in the first decade of the twenty-first century (UNESCO 2009; Obama 2010). At its heart, the conflict arises from regimes on either side of the border having diametrically opposed land-use ideas for the water in the shared watershed. In the United States (the downstream country), the transboundary Flathead is one of the most protected watersheds in the country. By contrast, it has been zoned for all manner of resource development in Canada. These largely incompatible regimes have created conflict between BC and Montana, provoked intense domestic debate within Canada, engaged the federal governments of the two countries, and triggered action under bilateral (the Boundary Waters Treaty) and multilateral international treaties (the World Heritage Convention). At the beginning of the twenty-first century, the conflict has steadily moved from a local and regional issue to one that has taken on an international profile.

In 2010, in front of the world press at the Vancouver Winter Olympics, the premier of British Columbia and the governor of Montana signed a memorandum of understanding (MOU) (Province of British Columbia and State of Montana 2010) in an effort to resolve this conflict. The MOU was followed by a discussion between the Canadian prime minister and the US president that acknowledged, but went beyond the MOU (Obama 2010). The existence of an MOU between a province and a state that purports to resolve international water issues raises clear questions relating to what levels of government have jurisdiction over international waters.

The MOU and subsequent actions by various actors in the Flathead also raise conceptual questions about equitable sharing of benefits, as well as questions about how affected parties engage in negotiation of international arrangements related to water governance that are intended to resolve conflicts. Despite the MOU, many issues central to the conflict remain unresolved at the time of writing.

Background

The Flathead River rises in the Canadian Rockies in the extreme southeast corner of BC and flows south into Montana, where it becomes known as the North Fork of the Flathead. The heavily forested valley has important ecological values, as well as significant coal resources. On the natural values side, it is a thoroughly studied, wild river with intact hydrological processes and healthy populations of endangered native fish, such as bull and cutthroat trout (Hauer and Muhlfeld 2010). The watershed is surrounded by wild lands with important biodiversity values, including the densest population of grizzly bears in the interior of North

America and a high concentration of vascular plants (Konstant et al. 2005), in addition to cherished recreational values. On the resource development side, the Canadian portion has long been known to contain significant coal deposits (Grieve and Kilby 1985), one of which underlies the river itself. The watershed is also subject to logging practices (both in Canada and the United States) and grizzly bear hunting in Canada. (See Figures 10.1 and 10.2.)

Figure 10.1 Photo of Flathead Valley.

Source: Harvey Locke.

Figure 10.2 Photo of Flathead River Valley.

Source: Harvey Locke.

The conflict over the management of the transboundary Flathead illustrates the basic tension that exists across the Yellowstone to Yukon region between conservation and resource extraction (Locke 2010). The Yellowstone to Yukon region contains some of the world's oldest and best-known national parks, and continues to be home to a wide range of carnivores and ungulates that have been extirpated elsewhere (Laliberte and Ripple 2004). This area also boasts a world-class native-trout fishery and is cherished for its wilderness and scenic values. However, this region also contains major coal, oil, hard rock minerals, and vast forests, some in globally significant concentrations, which continue to be attractive to extractive industries. These issues are especially compelling in the transboundary Flathead because of the fundamental tension between land use regimes and species-protection laws on the downstream US side, versus a public-policy regime that favours natural resource development and wildlife exploitation on the upstream Canadian side.

The vast majority of the US portion of the watershed is federal land, with one sizeable parcel of state land and some parcels of private land on the west bank of the river. Since 1910, the eastern half of the US river and valley has been protected by Glacier National Park. In 1976, Congress designated the entire US North Fork of the Flathead as a Wild and Scenic River to protect its environmental values further. These actions were also supplemented by strong national protections for fish and wildlife (US Fish and Wildlife Services 1973) and water withdrawal limitations to protect fluvial processes (State of Montana and Untied States 1993). In the Flathead National Forest, which occupies most of the west side of the US North Fork Basin, additional conservation measures control road densities and off-road vehicles. Indeed, the trend over the past 25 years in the US portion of the basin has been a steady increase in the number of policy measures designed to preserve and protect the environmental assets in the region (Sax and Keiter 2006). These conservation actions in the US Flathead are widely supported by an engaged public at the local level, where the area is called "The Gateway to Glacier National Park," and at the national level, where it has been called "America's Wildest Valley" (Robbins 2002). Figure 10.3 shows a map of the Flathead River Basin.

By contrast, the federal government of Canada has only recently expressed a formal desire to expand Waterton Lakes National Park into the Flathead Valley (Chrétien 2002) and has not acted to protect it – despite owning a significant parcel of land in the watershed: the Dominion Coal Block. Instead, it has deferred to the British Columbia provincial government. BC owns most of the Flathead Valley (subject to an unresolved land claim by the Ktunaxa First Nations) and has repeatedly zoned parts of the Canadian Flathead for

Figure 10.3 Map of the Flathead River Basin.

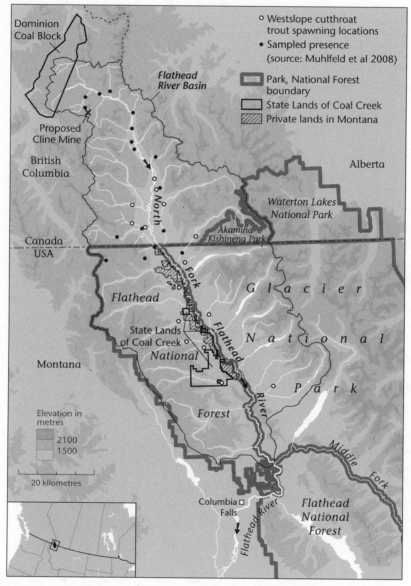

Source: Original Map by Eric Leinberger, University of British Columbia.

intense resource development, particularly coal mining (British Columbia Commission on Resources and the Environment 1994). This resource extraction – oriented regime has corresponded with a demographic situation in which the Canadian Flathead is very remote from population centres, little accessed, and largely unknown. Furthermore, to the north, it adjoins the Elk River Valley, which contains five open pit mines, making it one of the largest metallurgical coal-producing regions in the world with a culture of resource exploitation to match.

Forestry operations have taken place in the Canadian Flathead for many years with little public scrutiny, unlike the intense public interest that has been shown in forestry management and road densities on the US side. Grizzly bears are hunted in BC, but are highly protected on the US side. As this chapter is written, new areas continue to be logged and new roads cut because BC law offers roadless areas no protection under BC law. Road construction and associated logging carries the risk of adversely affecting the region's water quality, particularly by sediments impacting spawning fish. The only significant conservation designation was the 1993 creation of a small Akamina-Kishinena Provincial Park, which still allows hunting. The park is located in the headwaters of one tributary to the Flathead adjacent to Waterton Lakes National Park, which lies across the continental divide in neighbouring Alberta. In 2001, a significant part of the Flathead Valley in BC was declared a Wildlife Management Area for a brief period before an election, but this designation was immediately reversed by the subsequent government (British Columbia 2001, 2002). Though subject to resource exploitation, the Canadian Flathead remains very wild and without any human population. Until recently, the Flathead was one of southern Canada's least-known valleys, even to people living nearby.

The situation in the Canadian Flathead is particularly unusual given that it adjoins Waterton Lakes National Park (1895) in Canada and Glacier National Park (1910) in the United States, which were formally designated as Waterton-Glacier International Peace Park by acts of the Canadian Parliament and the US Congress in 1932. The conjoined federal parks became a World Heritage site under the World Heritage Convention in 1995, to which both Canada and the United States are signatories. Since the rise of large landscape conservation in the early 1990s, especially the Yellowstone to Yukon Conservation Initiative (Nature 2011), the transboundary Flathead has been the subject of great conservation and scientific interest (Locke 1994).

This attention is largely due to the region's important biodiversity values and its key position in the corridor between Waterton-Glacier International Peace

Park/World Heritage Site and the Canadian Rockies World Heritage Site cen-
tred on Banff National Park (Konstant et al. 2005; Locke 2010; UNESCO 2010).

The Boundary Waters Treaty

The BWT declares at Article IV that "waters flowing across the boundary shall
not be polluted on either side to the injury of health or property of the other"
(IJC 1909). As the basic legal framework for resolving conflicts between these
two nations over transboundary rivers, the BWT was used in the 1980s in an
effort to resolve the conflict over the transboundary Flathead.

In 1974, the Sage Creek Coal Company proposed a coal mine on the
banks of a small tributary of the Flathead called Cabin Creek. British
Columbia's government seemed receptive to the idea. It alarmed people
downstream in Montana. While Montana's Governor Ted Schwinden and
newly elected Congressman Max Baucus began meeting with the BC government, the Kalispell Lions Club and the Fernie (BC) Rod and Gun Club
also began to pay attention to this issue. This local, grassroots pressure
eventually produced petitions with tens of thousands of signatures in opposition to the mine (R. Moy, interview with M. McKinney, January 2010,
Helena, MT).

In 1982, a citizen-based environmental study concluded that the proposed mine posed a considerable threat to water quality (Flathead River Basin Environmental Impact Study, Steering Committee 1982). One year later,
the Montana legislature created the Flathead Basin Commission (1983) to
monitor and address water quality in the Flathead. Meanwhile, BC continued to review the mine proposal. After continued grassroots pressure and
some high-level diplomacy by US elected officials, both federal governments
agreed in 1984–1985 to use the Boundary Waters Treaty and refer the matter
to the IJC.

The IJC conducted an evaluation and issued its unanimous finding that
"the mine proposal as presently defined and understood not be approved"
(IJC 1988). The IJC further specified that the mine proposal should not receive regulatory approval in the future unless three conditions were fulfilled:
first, that the potential transboundary impacts the report identified were determined with reasonable certainty and would constitute a level of risk acceptable
to both federal governments; second, that the potential negative impacts on
sport fish populations and habitats in the Flathead River system would not occur or could be fully mitigated in an effective and assured manner; and third,
that the two national governments consider, with the appropriate jurisdictions,

opportunities for defining and implementing compatible, equitable, and sustainable development activities and management strategies in the upper Flathead River Basin. It is important to note the IJC's mention of fish and their habitats, as well as the direct implications for land use in this conclusion. Article IV of the BWT provides:

> It is further agreed that the waters herein defined as boundary waters and waters flowing across the boundary shall not be polluted on either side to the injury of health or property on the other.

Given that this article makes no mention of fish and their habitats or of land use, it begs a serious question – does the IJC's jurisdiction extend beyond water in the river channel? The IJC's condition that there be full mitigation of impacts on sport fish populations and their habitats is a concern derivative of water quality but not about water per se. Yet that did not deter the IJC, whose report went even further than considering fish and their habitats by recommending that the governments consider creating an International Conservation Reserve, an idea initially proposed by Montana Governor Schwinden during his testimony before them.

This IJC recommendation to create an International Conservation Reserve raises yet another point that has implications beyond the Flathead. Issues related to water quality across the border may not be resolvable without reference to the land-use activities that have a direct impact on that water and the many values it supports. As the profile of water as a precious natural asset with many values grows, we can expect that the BWT and other transboundary water agreements will be used in unexpected ways toward unexpected ends by people whose principle interests do not relate exclusively to water.

There may be nothing unusual about this situation within existing water governance frameworks. Remember that the primary role of the IJC is to resolve disputes over transboundary waters (see box 1.1 and chapter 4). For example, Article IX of the BWT provides:

> The International Joint Commission is authorized in each case so referred to examine into and report upon the facts and circumstances of the particular questions and matters referred, together with such conclusions and recommendations as may be appropriate, ...

As noted above, the IJC and all participants to the dispute over coal mining in the Flathead have long understood that the dispute, at its heart, is about not only the water in the channel (as is the case with irrigation or with drinking-water quality), but also the differing views of how to manage the watershed and attendant consequences for the natural values that water supports. Seen in this context, the IJC's recommendations regarding the protection of sport fish and their habitat, and the creation of an International Conservation Reserve, made perfect sense and could have been effective to resolve the dispute in the 1980s. But it fell on deaf ears.

Even though a high-profile BC politician, Davey Fulton, was a lead author of the report, the BC government never acknowledged or accepted the IJC recommendations. The problem seemed to go away when Sage Creek Coal Ltd voluntarily withdrew the Cabin Creek mining proposal in 1989. The IJC process garnered great attention in the United States during this dispute, but it failed to raise enduringly the Flathead's profile in Canada. That would change over time.

Resource Planning in British Columbia

During this same period, logging of coastal forests in British Columbia created significant public concern (Wilson 1998). Michael Harcourt and the New Democratic Party (NDP) won an election by promising a land-use plan for BC that would reconcile resource development and park creation. The Commission on Resource and the Environment (CORE) was set up to do regional plans, and the Flathead was in the East Kootenay Region Boundary. The public process recommended creating in the Flathead a significant provincial park that would cover the entire Kishinena Creek watershed above Glacier Park, Montana and adjacent to Waterton Park, Alberta (British Columbia Commission on Resources and Environment 1995). Efforts by Montana and Glacier National Park to incorporate the IJC-recommended International Conservation Reserve into this plan were unsuccessful. When the BC government ratified the plan, it decreased the large park's size to exclude the valuable forests in much of Kishinena Creek, leaving only the extreme high country abutting Waterton and Glacier Parks, one of which underlies the main stem of the Flathead River. It also called for expedited coal development in areas of coal deposits in the Flathead. BC's determination to treat the Flathead as an industrial area and disregard US concerns was clear. The lack of awareness and appreciation for the area in BC and Canada enabled the BC government's actions. In Canada, the Flathead was largely out of sight and out of mind. (See Figure 10.4.)

Figure 10.4 *Canadian Geographic* headline "Strip Mine This?"

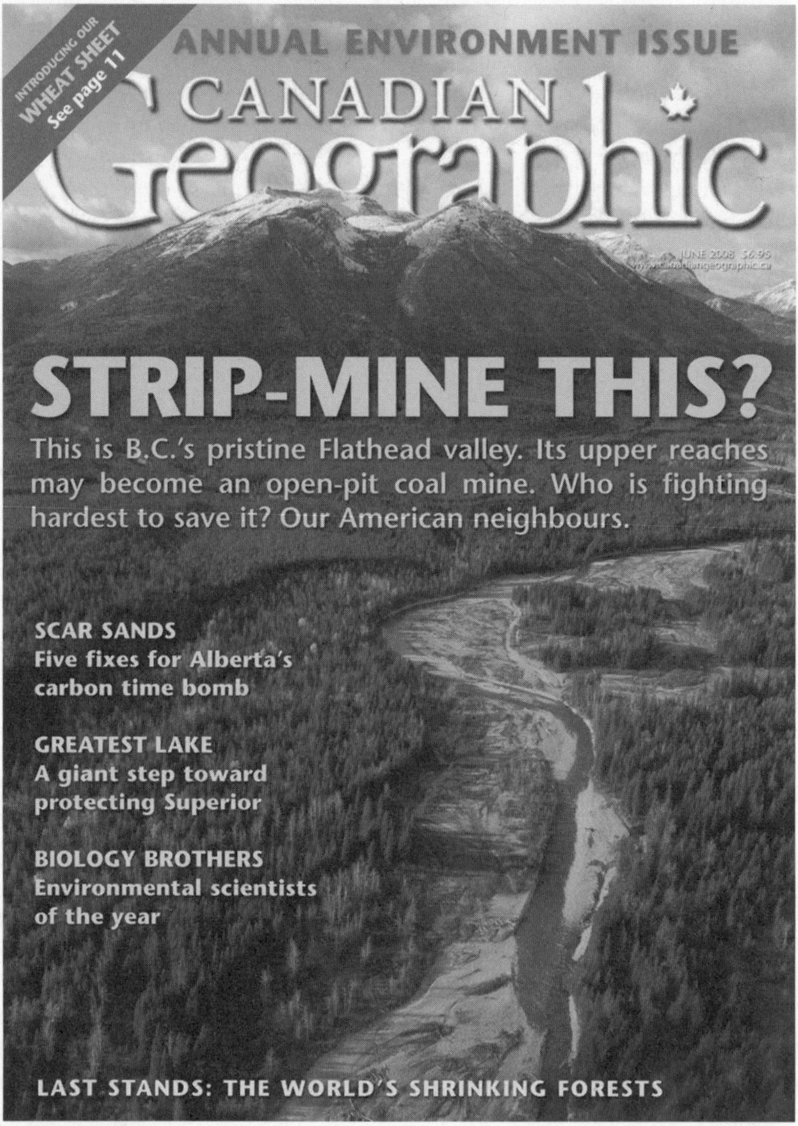

Source: Canadian Geographic.

An Abrupt Increase in Public Awareness

Public awareness in Canada about the transboundary Flathead has changed in the first decade of the twenty-first century. Canadians have become aware of the Flathead's ecological values due to a public awareness campaign by various nongovernmental organizations (NGOs), which worked together to create Flathead Wild. This coalition includes the Canadian Parks and Wilderness Society, Wildsight (an East Kootenay – based group), the Sierra Club of BC, the National Parks and Conservation Association, Headwaters Montana, and the Yellowstone to Yukon Conservation Initiative. The coalition also received strong scientific support from the Flathead Lake Biological Station of the University of Montana. Starting early in the new millennium, Flathead Wild worked hard to shift the management regime of the Canadian Flathead from resource extraction to conservation, which would roughly mirror the US side. They popularized the region's ecological riches and wilderness values, calling for the creation of a national park in one-third of the valley adjoining Waterton Glacier, as well as a Wildlife Management Area over the rest of the valley that would continue north to Banff National Park for wildlife connectivity (Flatheadwild ca. 2001). After an open-pit coal mine known as the Cline Mine was proposed in 2007 for the headwaters of the river in BC, Flathead Wild also called for a total ban on mining and oil-and-gas development in both the US and Canadian Flathead.

In early 2001, the BC NDP government under Premier Ujjal Dossanjh was receptive to Flathead Wild's proposals and became interested in reactivating the IJC idea of an International Conservation Reserve (John O'Riordan, interview with Harvey Locke, 2001). This government established the Southern Rockies Wildlife Management Area by an Order in Council (British Columbia 2001). Similarly, in 2002, the Prime Minister of Canada included the expansion of Waterton Lakes National Park into part of the Flathead Valley as an official part of the National Park System Plan for Canada, provided the province of BC and the Ktunaxa First Nation should agree (Chretien 2002). This was in response to a proposal from the Canadian Parks and Wilderness Society, an NGO that was very dissatisfied with the results of the CORE process. However, the BC NDP was voted out and replaced by a BC Liberal government led by Gordon Campbell, who campaigned against creating or expanding new parks. Instead, Campbell campaigned for establishing a two-zone mining policy, which meant that any area not already an existing park was open for mining. Not only was the park idea rejected, but the new government rescinded the Order in Council that had created the Wildlife Management Area and replaced it with the Southern Rockies Management Plan, which made mining a priority over all other uses (British Columbia 2003). Flathead Wild responded with an extensive public-awareness effort to increase knowledge of the natural values of the Flathead and their goals for the area.

A less well-known but equally important issue is that the Ktunaxa First Nations people identified the Flathead's importance to their culture through landclaim negotiations with the federal and BC governments. In 2004, the Ktunaxa formally asked the federal government to proceed with a national park feasibility study without taking a final position on whether they would agree to the creation of a park (Ktunaxa/Kinbasket Tribal Council 2004).

At the same time, Premier Campbell of BC and Governor Judy Martz of Montana signed the Environmental Cooperation Arrangement. The arrangement (*British Columbia – Montana Environmental Cooperation Agreement* 2004) calls for an "initiative to identify, coordinate and promote mutual efforts to ensure the protection, conservation and enhancement of our shared environment for the benefit of current and future generations" and states that the parties will develop an action plan within one year to achieve these mutual objectives. While this formal agreement did not specify the Flathead by name, it created additional hope in Montana for a transition from a resource development to conservation regime in BC. However, representatives from Montana walked away in 2006 after two years of negotiation, apparently frustrated by the lack of progress in shaping a plan that would effectively protect the transboundary Flathead.

Meanwhile, Flathead Wild's efforts to raise awareness were ironically aided in 2007 when the Cline Mine was proposed for the headwaters of the Canadian Flathead. This proposal shifted the public issue from one of a somewhat abstract campaign for a change of land-use regime from potential mining to active conservation, to one in which there was an immediate threat of environmental degradation to a pristine river from an open-pit mine.

A high-visibility conflict over the proposed Cline Mine ensued between the government of British Columbia and elected US officials, notably Governor Brian Schweitzer and Senators Max Baucus and Jon Tester. The dispute was featured in the news in both countries (*Canadian Geographic* 2008). Shortly thereafter, BC announced that it had recruited multinational energy company BP to develop coal-bed methane (CBM) in the Flathead and adjoining Elk River Valley. BC and Montana spent several years discussing the issue, but their differences were so fundamental that talks between them broke down. In an effort to protect against any suggestion of hypocrisy on the part of the Americans, supporters of Flathead Wild called for a ban on mining, oil, and gas development, as well as additional conservation measures on the US side of the basin through a multisignature letter to Senator Baucus (Christensen et al. 2009).

This attention focused largely on land use, which, of course, has implications for water. Indeed, the proposed coal mine and CBM led to the Flathead being declared "BC's Most Endangered River" on the front page of the *Vancouver Sun* (Pynn 2007). A cover story in *Canadian Geographic* (2008), with a photo of the beautiful Flathead River Valley, asked the pointed question, "Strip Mine This?" (See Figure 10.4).

The Conflict Becomes Global

To further raise the international awareness of the threat, Flathead Wild retained the nonprofit law firm Earthjustice to petition to have Waterton – Glacier International Peace Park declared a World Heritage Site in Danger by the United Nations Educational, Scientific, and Cultural Organization (UNESCO) (Earthjustice 2008). The petition, which enjoyed some support from the US National Park Service and Montana's Senators Baucus and Tester, was discussed at the 2009 World Heritage meetings in Seville, Spain; it raised concerns that were serious enough to warrant a fact-finding mission (UNESCO 2009). Thus, a second international treaty, the World Heritage Convention, came into play in the Flathead dispute. But unlike the BWT, which applies only to the United States and Canada, this convention is multilateral so the interests involved were international and efforts to solve the problem would have to consider this dimension also.

This international focus stirred the government of Canada into some public action in the form of Canadian Environment Minister Jim Prentice visiting the East Kootenay, and subsequent participation by Parks Canada in the World Heritage Mission to the region in September 2009. The week-long UNESCO mission garnered major press attention, and BC aggressively pled its case for responsible mining in the Flathead. But the IJC was never re-engaged to resolve the dispute by the national state parties.

Through the Flathead Wild campaign, the basin was highlighted by the International League of Conservation Photographers, and a high-profile hike garnered extensive front-page press in BC in the summer of 2009 (Pynn 2009; *Metro* 2009). Flathead Wild highlighted the Flathead in many ways at the World Wilderness Congress in Mexico in November of 2009 (which was attended by Canada's Minister of the Environment Prentice), and a resolution urging its protection was passed by the delegates who came from more than 50 countries (WILD 2009). The conflict escalated in early 2010 when it came to light that the UNESCO evaluation had determined that the Flathead would be a World Heritage Site in Danger should a mine proceed (Pynn 2010). This news was released just as international attention was being drawn to Canada as the host of the 2010 Winter Olympics in BC. The entire country risked embarrassment on the global stage over BC's management of the Flathead. But the federal government did not visibly intervene. Instead, the province resumed negotiations with Montana.

On the eve of the 2010 Winter Olympics, BC succumbed to the enormous domestic and international pressure by agreeing to remove the most egregious threat to the water and other ecological values of the Flathead through a change in the land-use regime. Premier Campbell announced a ban on mining, oil, and gas development in a Speech from the Throne in February (British Columbia 2010b). A few days later, he announced the MOU with Montana in the company of Governor Schweitzer. In the MOU, the governor promised to ban mining and oil-and-gas development on the US side of the border. But BC resisted the calls to protect the area's forests and wildlife with a national park and to restore the Wildlife Management Area. Instead, it reaffirmed that resource extraction of timber, quarrying, and wildlife exploitation would continue under its management regime. Particularly interesting and curious for many reasons was the substantial absence of the Canadian and US governments in the MOU and at the proceedings announcing it.

The Curious BC-Montana MOU of 2010

The BC-Montana MOU's four pages contain several points, but we will concentrate on the two most material provisions. The MOU states:

> British Columbia and Montana, the latter working with the United States as necessary, will implement measures necessary to prohibit the exploration for and development of mining, oil and gas, and coal in the British Columbia Flathead and the Montana North Fork Flathead River Basin, such action to be completed by July 2010, and subject to agreement on the equitable disposition of the financial implications of this action for the Province of British Columbia respecting existing mining and coal tenure holders.

In of refusing further conservation on the BC side, the MOU provided:

> Recognizing that the Flathead River Basin is the subject of uses that are important to local residents and that for approximately 70 years the British Columbia Flathead River Valley has been successfully managed for logging, recreation, guiding and outfitting, and trapping, that has maintained the healthy and diverse ecosystem that exists today.

The MOU removed a significant threat to downstream conservation values, but took no steps forward to enact conservation in the Canadian Flathead. This lack of positive conservation action failed to satisfy Flathead Wild and other interests in Montana. The detail of "equitable disposition of financial implications" also opened a Pandora's box of jurisdictional and political complexity that played out in a fascinating way over the next year.

Analysis of the Sub-national Effort to Resolve an International Issue

The BC-Montana MOU is curious for many reasons, not the least of which is its efforts to regulate foreign policy at the subnational level. Under both US and Canadian constitutional law, it is beyond question that foreign-policy and treaty-making authority lie with the federal parliament in Canada and with the president and senate in the United States. Both of these sovereign national governments have exercised that jurisdiction in relation to the transboundary Flathead through the IJC referral in the 1980s and the World Heritage nomination. No provincial- or state-level action can supersede these national actions. Many observers question whether state-province agreements can be anything more than nonbinding statements of intention. In any case, they certainly cannot bind federal land such as the Dominion Coal Block in BC or the vast amount of federal land in the US Flathead. Indeed, this subnational involvement speaks to the trends of rescaling identified elsewhere in this volume, particularly in Norman and Bakker's chapter 3.

These constitutional realities, however, did not deter Premier Campbell and Governor Schweitzer from signing the MOU without including the federal governments of the two countries. Remarkably, the Canadian government is not even mentioned in the MOU, and the US federal government is only mentioned derivatively, as follows: "Montana, the latter working with the United States as necessary, will implement measures necessary to prohibit the exploration for and development of mining, oil and gas, and coal in the British Columbia Flathead and the Montana North Fork Flathead River Basin."

The MOU largely ignored two other materially affected governments: the Confederated Salish and Kootenai Tribes and the Ktunaxa First Nation have a clear interest in the area, but were not included in the negotiations. At the last minute, representatives of these governments were invited to "witness" the signing of the MOU in Vancouver.

We can only understand the refusal to recognize the national governments that have international treaty-making authority by considering a broader context. An intense transboundary conflict related to softwood lumber and Devils

Lake vexed relations between the two countries in the preceding decade (see chapter 9, this volume, and Sax and Keiter 2006). To summarize briefly, Montana's Baucus was a vociferous opponent of Canadian lumber exports to protect the US timber industry in Montana. This earned him the enmity of the BC government, which presides over a significant timber industry whose major market is the United States. In addition, Schweitzer and Baucus, though both Democrats, do not enjoy a positive relationship. The governor excluded Baucus from the MOU negotiations, even though the senator had a demonstrated long-term interest in protecting the Flathead River. Furthermore, BC has sought to deny Canadian federal ownership of the Dominion Coal Blocks (Jessica McDonald, interview with Harvey Locke, 2003, Victoria, BC), despite its name and long-acknowledged federal ownership (Grieve and Kilby 1985). Finally, BC has not accepted the results of the 1980s IJC referral in the Flathead, so to use that mechanism again was unappealing to BC. The governor-to-premier dialogue that resulted in the MOU was a creative effort to address the conflict between BC and Montana given the politics of the day. However, it had built-in limitations.

BC promptly implemented the MOU through an Order in Council (British Columbia 2010b) and administrative action by the Gold Commissioner (British Columbia 2010a). These actions have the effect of law, but are easily revocable since they do not go through the legislature and have no process requirements. The governor returned to Montana with the MOU in hand and an image as the saviour of the Flathead. However, it was immediately obvious that the governor had very little ability to deliver unilaterally on the stated promises. The governor, in reality, had no authority over the 200,000 plus acres of existing oil-and-gas leases on the Flathead National Forest. Banning oil, gas, and mining development on the small parcel of state land on the US side of the border was all that the governor could do. The governor went before the State Board of Land Commissioners and asked for a ban on surface access to mines and minerals in the Coal Creek State Forest (Montana State Board of Land Commissioners 2010). What the governor got could be – at best – described as a start:

1. Subject to the requirements and restrictions ... of the U.S. Constitution, the ... Montana Constitution, and Section 77-5-116, MCA, the Board shall take steps to implement the goals of the February 18, 2010, MOU upon the Coal Creek State Forest and other State lands within the North Fork of the Flathead River Basin.
2. The Board shall impose a stipulation upon all mineral and oil and gas leases issued upon State lands within the Coal Creek State Forest and

within the North Fork of the Flathead River Basin. This stipulation will provide for no surface occupancy by any mineral lessee.

This action clearly did not extinguish any existing leases, nor did it fully ban oil-and-gas development or mining. Despite these shortcomings, the governor has made no further legislative or administrative efforts to fully implement the MOU on these Montana state lands.

It also became clear that the governor had promised money from the United States, not Montana, to BC to compensate the mining companies displaced by the BC ban. But the governor had no access to those financial resources, and he had not consulted his federal counterparts in advance.

Senators Baucus and Tester also had long-standing personal concerns about upstream mining in the Flathead. Understandably, they did not favour the idea of paying for the governor's promises, as discussed further below. They instead showed new interest in Flathead Wild's call for a federal ban on mining, oil, and gas development in the Flathead National Forest, which was also contemplated in the MOU. In May 2010, they introduced a bill to the US Senate to ban such activity and invited the Canadian ambassador to attend an evening reception in the senate building in Washington, DC, where the Flathead was feted. Senator Baucus's staff also worked very hard and successfully with several energy companies to ask them to voluntarily surrender leases in the North Fork that were still outstanding, but not developed, due to a previous judicial process. At the time of this writing, 80 per cent of such leases had been surrendered (Bureau of Land Management 2010).

To find permanent and effective solutions in the Canadian Flathead that reflected constitutional realities, Senator Baucus worked to include the White House (Baucus and Tester 2010). Similar efforts were made on the Canadian side to involve the federal government. These efforts bore fruit at the June 2010 G8/G20 summit meetings of world leaders in Canada when President Barak Obama and Prime Minister Stephen Harper found time to discuss the Flathead and issued a statement (Obama 2010). It noted the MOU, but more importantly, this statement recognized the need for more conservation efforts and federal involvement. It read:

On the margins of the Summit meetings in Canada this weekend, President Barack Obama and Prime Minister Stephen Harper noted the historic February 2010 memorandum of understanding between Premier Gordon Campbell of British Columbia and Governor Brian Schweitzer of Montana protecting the transboundary Flathead River Basin. They discussed how relevant US and Canadian agencies, including the US Department of the Interior and Environment Canada, working with representatives of the Province of British Columbia and the State of Montana, could support this understanding and could help ensure the sustained protection of the Flathead River Basin.

Thus, the heads of the two federal governments were now engaged, albeit belatedly, but not by engaging the IJC under the Boundary Waters Treaty. Instead, at least in part, they seemed to be approving, or at least acquiescing in, the subnational approach taken by BC and Montana,

The following month, in July 2010, UNESCO met and considered its mission report, the MOU, and BC's subsequent actions to ban oil, gas, and mining development. They noted that this positive development reduced the concern about a World Heritage Site in Danger, but also acknowledged the "ongoing threats to the property from possible impacts on wildlife connectivity arising from issues outside the property, including residential, industrial and infrastructure development, and forestry practices, in both Canada and the United States of America." Thus, wildlife connectivity between the Waterton-Glacier World Heritage Site and the Canadian Rockies World Heritage Site remained an outstanding issue that needed attention to secure the world-heritage values (UNESCO 2010).

Meanwhile, the July 2010 deadline on the MOU for "agreement on the equitable disposition of the financial implications" passed. It is useful here to examine the payment aspect of the MOU.

The Doctrine of Equitable Utilization and Sharing of Benefits

The emerging doctrine of equitable utilization in transboundary river disputes provides a useful analytical framework for the issues in the Flathead. Paisley (2002) describes this principle as the requirement that states act reasonably and equitably when dealing with transboundary water resources in their territory. This includes comparing the benefit derived from using the resource to the potential injury inflicted on the interests of another basin state. This principle in turn suggests the idea that there should be equitable sharing of benefits to the extent that a benefit occurs in one nation and costs are imposed in another.

The MOU between BC and Montana invokes some aspects of this principle. It provides that both parties will ban mining, oil, and gas development "subject to agreement on the equitable disposition of the financial implications of this action for the province of British Columbia respecting existing mining and coal tenure holders." It makes no provision to compensate the United States or Montana for having taken similar actions. This unilateral compensation aspect of the MOU may be explained by the principle of equitable sharing of benefits.

Montana wanted BC to forego mining, so it promised to solve BC's sense of obligation, legal or otherwise, to third-party mining companies adversely affected by such actions.

Absent from the MOU is any nonmonetary consideration of the benefits that BC and Canada's fish and wildlife have received because of the United States' significant conservation actions in the North Fork of the Flathead. These actions include forbearing certain kinds of resource extraction involving timber and wildlife, and investing in effective management regimes through the National Park Service, the US Forest Service, and the US Fish and Wildlife Service. Such benefits are easy to show since spatial maps of ecology clearly show that shared populations of fish and wildlife use both sides of the border seasonally (Muhlfeld et al. 2008; Weaver 2001). In an ironic twist, for many years the United States has protected grizzly bears at significant expense while BC kills them for the profit of guide outfitters who cater to US clients.

Canada and BC arguably have a duty to protect, or at least not harm, the conservation actions of their neighbour. The IJC stated in 1988 that "where one country has adopted uses with stringent environmental requirements in a boundary region on a unilateral basis that could preclude the otherwise legitimate development opportunities in the other, the parties should seek alternative development opportunities that are both sustainable and consistent with maintaining [treaty requirements] while paying regard to the legitimate goals of the other country" (IJC 1988).

No obvious reason in principle stipulates why the United States should pay BC to put in place partially comparable management of a shared resource. In the early 1990s, a strikingly similar series of events associated with the transboundary Tatshenshini – Alsek River system occurred where no compensation was paid. The two branches of the river rise in Canada (BC and Yukon), join and then flow into Glacier Bay National Park, Alaska. The headwaters of the Alsek River are protected in Kluane National Park Reserve, Yukon. A large open-pit copper mine was proposed for the unprotected headwaters of the Tatshenshini River, which rises in BC. In a high-visibility campaign, US Vice President Al Gore fought the proposed upstream Windy Craggy copper mine alongside Canadian and US nongovernmental environmental groups who called themselves Tatshenshini Wild. The BC government, led by NDP Premier Harcourt, responded favourably to the alternate land designation of a park proposed by Tathsenshini Wild. BC, without any US financial contribution, paid compensation to the owner of the proposed mine and declared, on establishing the new park, that it was "British Columbia's gift to the world." In the case of the Flathead, an apparent political and pragmatic reason for this payment achieved less compatible alignment of land-use designations than in the Tatshenhini case;

BC apparently needed some political and financial help in achieving its goal, and the governor was prepared to give it, aided by the hope that the funding would come from someone else.

The Montana senators showed very little interest in using the US federal government to pay BC the money the governor had promised to the premier (Baucus and Tester 2010). It seems understandable that they would not want to pay the bill for a lunch to which they were not invited, so to speak. Matters were exacerbated by the deteriorating US fiscal condition, which was a major electoral issue in 2010. Furthermore, BC had only implemented the ban by Order in Council and Administrative Activity, not by a special law. Knowledgeable observers therefore had little confidence in the permanence of the designations, which could be undone as quickly as they had been implemented and without any public process. Indeed, the same BC government headed by Campbell had rescinded the earlier Wildlife Management Area Order in Council in the Flathead immediately on taking office (British Columbia 2002). But the governor did not view the situation that way. He spoke out publicly about how the senators had let him down and thus put the Flathead at risk (Gouras 2010), stating that they needed to come up with $16 million USD to pay for the deal. This very public conflict caused a major stir among Montana Democrats and provoked at least one newspaper editorial that sharply criticized the governor (*Helena Independent Record* 2010).

Ultimately, the political difficulty of this issue grew for both the governor and the two senators. Secretary of the Interior Ken Salazar and Canadian Ambassador Gary Doer were also engaged. They all agreed that the ban on oil-and-gas development in BC should remain in place. Noting the limited financial and political ability of the US government to pay compensation, they appealed to nongovernmental organizations, namely The Nature Conservancy (TNC) in the United States and the Nature Conservancy of Canada (NCC), which are arm's length organizations, despite the similarity of their names. The NCC had access to Canadian federal money for habitat and connectivity conservation. Months of negotiations involving various parties followed, the full participants and details of which are unavailable. Then, in February 2011, BC's Speech from the Throne (British Columbia 2011a), just before Campbell was set to step down as premier, stated, "We look forward to implementing this agreement with our partners and paralleling steps taken recently in the United States Congress and by the State of Montana." Everyone understood this announcement to mean that BC would implement the ban on oil, gas, and mining development legislatively in return for $9.4 million to be paid by TNC and NCC.

The actual agreement between BC, TNC, and NCC remains unavailable to third parties at the time of writing. Therefore, information on this subject must come from press releases. The premier of BC's press release stated that the provincial government "will introduce legislation to support the 2010 MOU" and that "Under the Agreement on the Protection of the Transboundary Flathead Watershed Area, the [NCC] and [TNC] will contribute $9.4 million to implement the environmental protection provisions of the MOU, including compensation for current coal and mineral tenure holders for their past exploration." It further indicates that TNC would raise the funds from private sources and NCC would pay for its share with funds it had received from the federal government of Canada's Natural Areas Conservation Program. Thus, in a remarkable and mind-bending turn of events, Canada's federal government became a major funder of the MOU between BC and Montana to pay for part of the governor's promise (British Columbia 2011b).

The Backgrounder to the Press Release states that "under the Agreement ... the Province undertakes to continue its best efforts to implement the MOU," including measures to "introduce a bill to legislate the legal and regulatory measures the province took in February 2010 to prohibit exploration for and development of mining, oil and gas and coal in the BC Flathead." The BC press release also said, "This agreement will ensure that the healthy ecosystem that exists today in the Flathead River Basin will continue to be maintained in a manner consistent with current recreation, forestry, guide outfitting and trapping uses." The Backgrounder also provides that, subject to funding, there will be accelerated designation of Wildlife Habitat Areas for bull and cutthroat trout and some other species. Notably, grizzly bears are absent from this list.

The BC did not mention satisfying UNESCO's concerns about connectivity. However, a simultaneous press release by Secretary of the Interior Salazar implies that the payments could open the door to further conservation actions in BC (Salazar 2011). This press release was accompanied by a reception in Washington, DC, which was attended by Canadian Ambassador Doer, as well as representatives of TNC and NCC. After noting the MOU and subsequent BC-TNC-NCC deal, the press release concluded:

In June 2010, President Barack Obama and Prime Minister Stephen Harper committed to cooperate on sustainable protections in the Flathead Basin. Since then, the US Department of State and the Canadian Embassy have conducted extensive diplomatic outreach over the past eight months to bring together Federal, provincial, and state government stakeholders to identify common interests and steps toward protection in the Flathead River Basin.

The Department of the Interior will continue to coordinate with its partners to advance permanent, sustainable protections in the Flathead watershed, including the State of Montana, the Confederated Salish and

Kootenai Tribes, British Columbia, the Ktunaxa First Nation, and Environment Canada – including Parks Canada and the Canadian Wildlife Service – through the Great Northern Landscape Conservation Cooperative. This framework will be used for further engagement in the months ahead, including to conduct joint activities on fish and wildlife conservation, invasive species and pests, climate change adaptation, and environmental data collection and information sharing.

This hint of other conservation actions to follow through the Great Northern Landscape Conservation Cooperative opens the potential of further federal involvement. The idea of a Landscape Conservation Cooperative falls under the America's Great Outdoors (AGO) Initiative of US President Barack Obama; it calls for large landscape conservation cooperation, both among US agencies and with Canada and Mexico at the scale of Yellowstone to Yukon (Council on Environmental Quality et al. 2011). The transboundary Flathead is an obvious area of potential interest (Locke 2011–12). Similarly, in May 2011, Prime Minister Harper was re-elected with a majority. A piece of his platform calls for a National Conservation Plan for Canada (Conservative Party of Canada 2011), which could be similar to President Obama's AGO Initiative. The Parliamentary Standing Committee on Environment and Sustainable Development held hearings on the National Conservation Plan in May 2012 but the outcome is not clear at the time of writing.

Meanwhile, British Columbia politics changed. Premier Campbell resigned. There was concern whether his successor, Christie Clark, would keep the commitment Campbell had made to legislate a ban on mining and oil and gas. She in fact did follow through. The October 2011press release (British Columbia 2011c) announcing that the legislation indicates how her government viewed this action as tightly connected to broader land-use issues in the watershed, as opposed to exclusively water governance issues: "VICTORIA – The Flathead Watershed Area Conservation Act, introduced Oct. 4, 2011, is intended to preserve the environmental values in the Flathead watershed.

The Flathead, in the East Kootenays, neighbours the Waterton Glacier International Peace Park and is listed by UNESCO as a World Heritage Site and Biosphere Reserve. The introduction of legislation meets the commitment made in the 2010 MOU signed with the state of Montana on Environmental Protection, Climate Action and Energy. The act, when passed and brought into force, will secure decisions made in 2010 to:

- Establish coal and mineral reserves.
- Prohibit Crown land dispositions for mining purposes.
- Prohibit issuance of Mines Act permits.

- Prohibit issuance of Oil and Gas Activities Act permits for oil and gas exploration and development.
- Prohibit disposition of Crown reserves under the Petroleum and Natural Gas Act.

Parallel legislation has been introduced in the US senate to similarly remove mining, oil, and gas as permissible land uses in the Montana North Fork Flathead Basin, which neighbours B.C.'s Flathead valley. More than 80 per cent of existing oil-and-gas leases have already been retired in the area in the United States."

The lead item under Quick Facts in the Press Release noted: "The Flathead River is one of North America's last wild rivers, and the area supports a variety of animals and a diverse collection of plants and fish." The Flathead Watershed Conservation Act passed shortly thereafter in November 2011 (British Columbia 2011d). It is noteworthy that the Act's title focuses on the watershed and its conservation, not just the river or water. Thus, over 30 years of intense conflict and efforts at resolution, the basic views of land use along the Flathead River on both sides of the border are starting to converge.

Conclusions

Short-term questions and long-term lessons arise from the latest twists in the story of the transboundary Flathead. BC followed through on its commitment to legislate an end to oil-and-gas mining in the Flathead. In so doing, it brought the land-use regimes on the two sides of the border into closer alignment and thus removed some of the basic tension that led to the engagement of the BWT and the IJC in the Flathead and the subsequent events that did not involve them. Ironically, at the time of writing, this alignment of land use now requires further action on the US side and from the government of Canada. This situation show the issues are not all resolved.

There is doubt about whether Bill 233 in the US Senate (2011) to ban oil, gas, and mining development on the US side will pass. What will BC do if it does not? Similarly, Governor Schweitzer made progress on ending oil-and-gas mining on lands in Montana by banning surface access, but unlike his BC counterparts, the Montana governor did not legislate an end to it. What would BC or Canada do if a subsequent government of Montana chose to allow access to coal, oil and gas, and minerals from adjoining parcels in its part of the Flathead watershed, which is downstream from Canada but that would likely affect the upstream fisheries? There is also the Dominion Coal Block, a parcel of federal land in the Canadian Flathead, which has no legislative framework banning oil and gas that applies to it. What will happen there?

Two other questions arise. The land-use decision implemented in BC does not satisfy all actors. Will other land-use decisions be made that would help to end the dispute permanently? What about the mutually compatible national conservation efforts announced by President Obama and the potentially similar plan promised by Prime Minister Harper? Will UNESCO's unaddressed concerns about connectivity raise further World Heritage concerns and further engagement of the international community? Equally unknown is how the Ktunaxa Nation treaty process will affect land use on the Canadian side. Flathead Wild also seeks further conservation on the US side in the form of wilderness designations. Will those efforts succeed?

Flathead Wild has signalled its clear intention to press on with its efforts in Canada, the United States, and abroad to have Waterton Lakes National Park expanded into the Flathead and for a Wildlife Management Area to be restored to provide connectivity. Will a future BC government embrace these goals?

Regardless of the uncertainties facing the Flathead watershed, this case study suggests take-away lessons that might inform and invigorate similar transboundary resource negotiations. First, it is critically important to involve all affected parties in the negotiation as early as possible. National governments are key players in transboundary problem-solving. However, as seen in this case and others in this volume, it is increasingly clear that the inclusion of parties, as well as federal governments, is essential to the success of the long-term governance of shared waters. In this case, it was important to engage state, provincial, and indigenous governments, as well as nongovernmental actors. But, there is thus no consensus among conservation NGOs that an effective land-use decision has been reached. Providing meaningful opportunities to engage stakeholders, policymakers, and administrators is fundamental to building the sense of ownership that is needed to implement any negotiated agreement effectively.

Finally, this case study illustrates the value of engaging or taking advantage of third parties and events – in this case, the International Joint Commission, UNESCO, and the Winter Olympic Games. Third parties help raise visibility and awareness, clarify and sharpen issues and options, and galvanize action. They are key actors in land-use decisions and water governance.

It is clear that for so long as the land-use regimes in the watershed do not closely resemble each other across the Canada- US border, the issues connected to the Flathead River are likely to remain highly visible for some time.

REFERENCES

Baucus, Max, and Jon Tester. 2010. Senator's letter to Governor Schweitzer on North Fork Protection, 30 June 2010. Washington, D.C.: Office of Senator Baucus. British

Columbia. 2001. *Southern Rocky Mountain Conservation Order*. Order in Council 428. Victoria.

British Columbia. 2002. *The Southern Rocky Mountain Conservation Order is Repealed*. Order in Council 246. Victoria.

British Columbia. 2003. *Southern Rocky Mountain Management Plan*. Victoria, British Columbia.

British Columbia. 2010a. *Regulation of the Gold Commissioner*. 4 February, BC Reg 41/2010. Victoria.

British Columbia. 2010b. "Speech from the Throne." 9 February. Hansard, Victoria.

British Columbia. 2011a. "Speech from the Throne." 14 February. Hansard, Victoria.

British Columbia. 2011b. "BC, Montana and Partners Unite to Sustain Flathead" *Office of the Premier* 2011. PREM 0011 000138. 14 February. Victoria.

British Columbia. 2011c. http://www.newsroom.gov.bc.ca/2011/10/legislation-introduced-to-protect-flathead-watershed.html. 4 October 2011, clarified 5 October 2011, Victoria.

British Columbia. 2011d. Flathead Watershed Conservation Act, SBC 2011, c. 20 British Columbia, Ministry of Sustainable Resource Management, 2003. *Southern Rocky Mountain Management Plan* Objective 3.1.2. Victoria, BC.

British Columbia Commission on Resources and the Environment. 1994. "The East Kootenay Boundary Plan." October, Victoria.

British Columbia Commission on Resources and the Environment. 1995. "The East Kootenay Land Use Plan." Victoria.

British Columbia-Montana Environmental Cooperation Agreement. 2004. http://www.env.gov.bc.ca/spd/ecc/docs/bcwaccord.pdf.

Bureau of Land Management. 2010. North Fork of the Flathead Lease Surrender Map.

Canadian Geographic. 2008. "Strip Mine This?" Annual Environment Issue, June, Ottawa.

Chrétien, Jean. 2002. "The Government of Canada Announces Plans to Protect Canada's Natural Heritage." News Release CP2002 000469, 3 October, Ottawa.

Christensen, Dana, et al. 2009. Group letter to Senator Baucus re recommended conservation actions in US North Fork of Flathead, 3 July, Kalispell, Montana.

Conservative Party of Canada. 2011. "Here for Canada: Stephen Harper's Low-Tax Plan for Jobs and Economic Growth." Ottawa.

Council on Environmental Quality et al. 2011. "America's Great Outdoors: A Promise to Future Generations: Executive Summary." Washington D.C.

Earthjustice. 2008. Petition Letter to UNESCO to Declare Waterton-Glacier International Peace Park a World Heritage Site in Danger, June, San Francisco.

Flathead Basin Commission. 1983. "Purpose." MCA 75–7–302. http://data.opi.mt.gov/bills/mca/75/7/75-7-302.htm.

Flathead River Basin Environmental Impact Study, Steering Committee. 1982. "Final Report of the Steering Committee for the Flathead River Basin Environmental Impact Study."

Gouras, Matt. 2010. "Schweitzer Says Federal Government 'Let Us Down' on Protecting North Fork of the Flathead." *Missoulian*, 4 June.

Grieve, D.A., and W.E. Kilby. 1985. "Flathead Ridge Coal Area Southern Dominion Coal Block Parcel 82, Southern British Columbia." British Columbia Ministry of Energy Mines and Petroleum Resources Geological Fieldwork Paper 1986–1. http://www.em.gov.bc.ca/Mining/Geoscience/PublicationsCatalogue/Fieldwork/Documents/1985/05_grieve_p25-36.pdf

Hauer, F.R., and Clint C. Muhlfeld. 26 Mar, 2010. "Compelling Science Saves a River Valley." *Science* 327 (5973): 1576. http://dx.doi.org/10.1126/science.327.5973.1576-a. Medline:20339049

Helena Independent Record. 2010. "Governor's Grandstanding Wearing Thin." Editorial 13 June. Helena, Montana.

International Joint Commission (IJC). 1909. "Boundary Waters Treaty Text." http://bwt.ijc.org/index.php?page=Treaty-Text&hl=eng.

International Joint Commission (IJC). 1988. "Impacts of a Proposed Coal Mine in the Flathead River Basin." International Joint Commission, December. http://www.ijc.org/php/publications/pdf/ID590.pdf.

Konstant, W., H. Locke, and J. Hanna. 2005. "Waterton-Glacier International Peace Park: The First of its Kind." In *Transboundary Conservation: A New Vision for Protected Areas*, ed. R.A. Mittermeir et al., 70–6. Mexico: Cemex-Agrupacion Sierra Madre – Conservation International.

Ktunaxa/Kinbasket Tribal Council. 2004. "Letter to Honourable Stephan Dion, Minister of the Environment Regarding Proposed Expansion of Waterton Lakes National Park." 29 November, Cranbrook, BC.

Laliberte, Anrea S., and William J. Ripple. 2004. "Range Contractions of North American Carnivores and Ungulates." *Bioscience* 54 (2): 123–38. http://dx.doi.org/10.1641/0006-3568(2004)054[0123:RCONAC]2.0.CO;2.

Locke, Harvey. 1994. "The Wildlands Project and the Yellowstone to Yukon Biodiversity Strategy." In *Borealis* 15 (Winter). Ottawa: Canadian Parks and Wilderness Society.

Locke, Harvey. 2010. "Yellowstone to Yukon Connectivity Conservation Initiative." In *Connectivity Conservation Management a Global Guide*, ed. G. Worboys, W.L. Francis, and M. Lockwood, 161–81. London: Earthscan.

Locke, Harvey. 2011–12. "Transboundary Cooperation to Achieve Wilderness Protection and Large Landscape Conservation," *Park Science* 28 (3). National Park Service, US Department of Interior.

Metro. 2009. "Metro Flathead RAVE." 5 August. Vancouver, BC.

Montana State Board of Land Commissioners. 2010. "Resolution of the State Board of Land Commissioners To Implement Restrictions on State Lands Located in the North Fork of the Flathead River Basin." 18 March.

Muhlfeld, C., M. Deleray, and A. Steed. 2008. "Canadian Energy Development Threats and Native Fish Research and Monitoring in the Transboundary Flathead River System." Unpublished document, Montana.

Nature. 2011 "Think Big." *Nature Editorial* 469 (131), 13 January.

Obama, Barack. 2010. "Statement by the Press Secretary on Protecting the Flathead River Basin." The White House, Washington D.C., 28 June.

Paisley, Richard. 2002. "Adversaries into Partners: International Water Law and the Equitable Sharing of Downstream Benefits." *Melbourne Journal of International Law* 3:280–300.

Province of British Columbia and State of Montana. 2010. "Memorandum of Understanding and Cooperation on Environmental Protection, Climate Action and Energy between the Province of British Columbia and the Governor of the State of Montana." 18 February.

Pynn, Larry. 2007. "Flathead River Makes Most Endangered List." *Vancouver Sun*, 12 March.

Pynn, Larry. 2009. "Time for Peace in the Flathead Valley." *Vancouver Sun*, 1 August.

Pynn, Larry. 2010. "UN Seeks Flathead Valley Mining Moratorium." *Vancouver Sun*, 22 January.

Robbins, Jim. 2002. "Where the Bears and Wolverine Prey." *New York Times*, 16 July.

Salazar, Ken. 2011. "Secretary Salazar Joins Canadian Ambassador Doer in Celebrating Agreement to Protect Transboundary Flathead River Basin." Department of the Interior Press Release, 15 February, Washington, D.C.

Sax, Joseph L., and R.B. Keiter. 2006. "The Realities of Resource Management: Glacier National Park and its Neighbours Revisited." *Ecology Law Quarterly* 33 (2): 233–312.

Senate (US) Bill 233. 2011. *The North Fork Watershed Protection Act of 2011.*

State of Montana and United States. 1993. *Water Rights Compact, State of Montana, United States of America, National Park Service.* 85–20–401, USC and 85–20–401, MCA. http://data.opi.mt.gov/bills/mca/85/20/85-20-401.htm.

United Nations Educational, Scientific and Cultural Organization. 2009. "Waterton-Glacier International Peace Park (Canada/US) (N354 rev)." Report of Decisions WHC-09/33 COM 20 p 71. Seville, Spain.

United Nations Educational, Scientific and Cultural Organization. 2010. "Waterton-Glacier International Peace Park (Canada/US) (N 354rev)." Decision – 34COM 7B.20 – Waterton-Glacier International Peace Park (Canada / United States of America). Brasilia, Brazil.US Fish and Wildlife Services. 1973. *Endangered Species Act.* 7 U.S.C. § 136.

Weaver, John. 2001. "The Transboundary Flathead: A Critical Landscape for Carnivores in the Rocky Mountains." Wildlife Conservation Society Working Paper 18, New York.

WILD. 2009. "Resolution 5: Flathead National Park and Canadian Rocky Mountain Wildlife Connectivity/Conectividad entre el Parque Nacional Flathead y la vida silvestre en las montañas Rocallosas canadienses." The WILD Foundation. http://www.wild.org/resolutions-from-wild9.

Wilson, Jeremy. 1998. *Talk and Log: Wilderness Politics in British Columbia*. Vancouver: University of British Columbia Press.

11 The Great Lakes: A Model of Transboundary Cooperation

JAMIE LINTON AND NOAH HALL

The Canadian wilderness was white with snow. From Lake Superior northward the evergreen trees wore hoods and coats of white. A heavy blanket of cloud hung low across the hills. There was no sound. Nothing moved ...

So begins Holling Clancy Holling's classic children's book, *Paddle to the Sea,* which describes the voyage of a small carved canoe, beginning on a river entering Lake Superior and travelling through each of the five Great Lakes to the St. Lawrence River and eventually to the Atlantic Ocean (Holling 1941). The book, written and illustrated by an American, was first published in 1941 and remains in print. A film of the same title based on Holling's book was produced by the National Film Board of Canada in 1966 and directed by the famous Canadian naturalist and author, Bill Mason. *Paddle to the Sea* is a fitting introduction to this chapter: the book and the film have educated generations of children on both sides of the border about the physical, economic, and cultural dimensions of a distinct place known as the Great Lakes. Like the Great Lakes, *Paddle to the Sea* is both Canadian and American, and has helped to produce a unique history and culture. The unfolding of that history, in turn, has influenced the Great Lakes, contributing to a process where nature and society have evolved in relation to one another (Linton 2010, 24–44). In this respect, this chapter is different from the others in Part II of this volume: rather than seeing it as a "flashpoint," we see the Great Lakes as a transboundary success story. For the most part, the lakes have given Canadians and Americans opportunities to collaborate.

The Great Lakes continue to be significant in different ways to Americans and Canadians. The lakes constantly present both countries with new challenges, and they have sometimes acted as a source of contention between the

Figure 11.1 Map of the Great Lakes Basin.

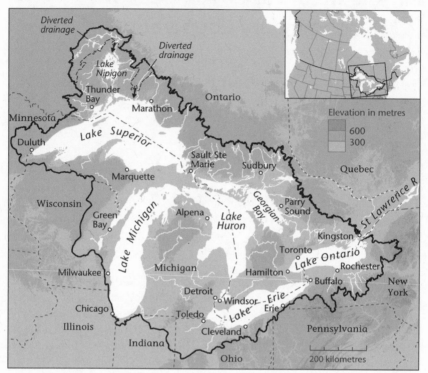

Source: Original Map by Eric Leinberger, University of British Columbia.

two countries. However, the thesis argued in this chapter is that these challenges have been rendered into opportunities for cooperation and collaboration: the constant emergence of new problems and issues is balanced by a history of commitment to overcoming such challenges together. Figure 11.1 shows a map of the Great Lakes Basin.

Physical and Cultural Background

The size of the Great Lakes is such that they invite ready comparison with the sea, and indeed, they are often described as "inland seas" and as forming the "third (or fourth) coast." The size of the Great Lakes alone makes them a pivotal player in the continent's history. They contain more than 80 per cent of North America's surface water, and together, they comprise the largest readily available source of freshwater on Earth, boasting 21 per cent of the world's surface water (Environmental Protection Agency 2011b). The surface area of all five lakes nearly equals that of the United Kingdom, and their combined shoreline measures over 17,000 km (10,500 miles), nearly half the circumference of the Earth (Grady 2007, 21). These lakes are also incredibly ecologically diverse, providing habitat for more than 350 species of fish and 3,500 species of plants and animals (International Joint Commission 2006, iv).

The Great Lakes and St. Lawrence River form a single hydrological system linking the interior of North America with the Atlantic Ocean. Starting with a river flowing south into Lake Superior from the Nipigon region, this system is described by Holling as follows:

> The river flows into the Great Lakes, the biggest lakes in the world. They are set like bowls on a gentle slope. The water from our river flows into the top one, drops into the next, and on to the others. Then it makes a river again, a river that flows to the Big Salt Water. (Holling 1941, 3)

At the upper end of the system, the surface of Lake Superior averages about seven metres (24 feet) above that of Lakes Huron and Michigan, which form a single hydrological unit. While the water level in Lake Erie is only a few metres lower than in Lakes Huron – Michigan, the drop from Lake Erie to Lake Ontario (over the Niagara Escarpment at Niagara Falls) is almost 100 metres. The amount of water flowing through this system is a tiny proportion of the total volume of the lakes themselves: less than 1 per cent per year. The hydrology of the Great Lakes has important social implications. For example, since the lakes form a single hydrological system, in some respects, the whole system needs to be managed in an integrated fashion. The outflow of the lakes also has social implications. Outflow is so small that pollutants entering the system tend to persist and become more concentrated with time (Environmental Protection Agency 2011b). In addition, the relatively small proportion of renewable water flowing in the system means that little or no "surplus" water is available for removal from the basin. The various jurisdictions sharing these waters have therefore need to work together to protect the Great Lakes from excessive consumptive uses (International Joint Commission 2000).

The Great Lakes have both separated and joined the United States and Canada, serving as a foundation for international trade, as well as cultural contact and exchange. In the early seventeenth century, during initial European exploration of the region, the lakes were bordered by two major linguistic groups (Algonquian and Iroquoian) and dozens of distinct nations (Heidenreich and Wright 1987). Today, the international significance of the Great Lakes is suggested by the fact that they define a large portion – about a third – of the border between the United States and Canada. The border itself bisects Lakes Superior, Huron, Erie, and Ontario. Lake Michigan is the only Great Lake that is entirely within the boundary of one country: the United States. The province of Ontario occupies the entire Canadian side of this border, and the US side includes eight states: Minnesota, Wisconsin, Illinois, Indiana, Michigan, Ohio, Pennsylvania, and New York. When combined with the hydrological imperative of integrated management, the international and interjurisdictional complexity of the region presents governance challenges that people living in the basin have had to overcome.

The Great Lakes Basin is small relative to the size of the lakes themselves; nevertheless, it occupies an area approximately the size of France and the UK combined. Some 40 million people live in the Great Lakes Basin, the vast majority of which reside in the southern area around Lakes Erie and Ontario, and the southern portions of Lakes Michigan and Huron (Grady 2007, 32). Americans make up about three-quarters of this population with roughly 30 million people, and Canadians comprise about 10 million. Despite the immense capacity and historical resilience of the Great Lakes, excessive withdrawals and prolonged consumptive use by these 40 million human inhabitants could negatively impact water levels and the environment. However, despite recent population growth, total water withdrawals from the Great Lakes have decreased or increased only slightly most years since 1990, most likely due to conservation and more efficient water use (Reeves 2011, 28–9). The most significant anthropogenic uses of water in the basin, in order of volumes withdrawn, are as follows: cooling thermoelectric power plants, providing public water supply, industrial use, and irrigation. Along this portion of the shared border, roughly 23 million people depend on the Great Lakes for their drinking water (Sproule-Jones 2002, 3).

Perhaps drinking from the same source of water produces a kind of conviviality among people. In any case, citizens living along the shores and in the basin of the Great Lakes evidently share elements of a common culture. This common culture is reflected in the colourful title of Ted McClelland's popular 2008 book, *The Third Coast: Sailors, Strippers, Fishermen, Folksingers, Long-Haired Ojibway Painters, and God-Save-the-Queen Monarchists of the Great Lakes*. Mc-Clelland sets the scene for his stories of people living in the basin by declaring: "The Great Lakes are the North's most remarkable natural feature ... They also comprise a nation within North America, as surely as the South. I call it the Freshwater Nation" (McClelland 2008, xiii).

This "Freshwater Nation" nevertheless has different histories and inhabitants on either side of the international border experience it somewhat differently, a dynamic that is explored in the following two sections.

The Great Lakes in Canadian History, Economy, and Culture

The article on the Great Lakes in *The Canadian Encyclopedia* frames the subject in the following terms: "[The] Great Lakes are the largest group in a chain of large lakes (including Winnipeg, Athabasca, Great Slave and Great Bear) that lies along the southern boundary of the Canadian Shield" (Marsh 1985, 934). While geologically correct, this framing reflects the rather proprietary attitude that many Canadians take toward the Great Lakes. From a Canadian perspective, it is something of an inconvenience to share the lakes with their US neighbours. Indeed, an early Canadian geography textbook found it necessary to point out, "Only part of the Great Lakes belongs to Canada" (Morrison 1930, 493).

This Canadian sense of ownership of the Great Lakes is partly explained by their relative importance in Canada's history. The Great Lakes figure in the founding stories of Canadian nationhood, beginning with the travels of Samuel de Champlain in the early seventeenth century and continuing with the continental expansion of the St. Lawrence fur trade over the following two centuries. Traders used the lakes as means of transport, sites for trading, and entrepôts, eventually fanning out from the Great Lakes to bring much of the continent into a system of trade. In the nineteenth century, the development of Canada as a nation and a national economy along an east-west axis hinged on the Great Lakes and the St. Lawrence River. This hydrological system served as a commercial artery and communications system linking the Atlantic provinces with Manitoba, Saskatchewan, Alberta, and eventually British Columbia on the Pacific Coast (Easterbrook and Aitken 1956).

As Canada's economic centre of power shifted westward from Montreal to Toronto during the course of the twentieth century, the Great Lakes region assumed increasing national importance. To the extent that the "view from Toronto" has attained a measure of hegemony in defining and shaping Canadian identity, the Great Lakes have assumed a central place in that identity. Today, the population of the Great Lakes Basin is comprised of one third of Canada's population, compared with just 10 per cent of the US population.

The Great Lakes also figure prominently in the Canadian arts. Thomas Cary's "Abraham's Plains," one of the earliest and most anthologized poems written in Canada, put the Great Lakes at the centre of the Canadian landscape (Bentley n.d.). Gordon Lightfoot's 1976 hit, "The Wreck of the Edmund Fitzgerald," commemorated the famous 1975 sinking of the iron-ore carrier of that name on Lake Superior and has been described as "perhaps the most well known Canadian song about water" (Environment Canada 2001). This demonstrates the importance of the Great Lakes in Canadian music, as does the current popularity of the indie-pop group called Great Lake Swimmers and adoptive Canadian composer Hennie Bekker's *Great Lakes Suite*. These lakes have an important place in the visual arts, too: Lauren Harris's iconic paintings of the north shore of Lake Superior contribute to the oeuvre of the Group of Seven, whose distinctive early-twentieth-century style was inspired by the landscapes of the Great Lakes Basin, especially Georgian Bay, Muskoka, Algoma, and Algonquin Park.

The Canadian public's deep concerns about exporting water in bulk are partly a reflection of the sense of ownership and importance of the Great Lakes to Canadian identity. The prospect of bulk water exports is a longstanding issue of concern to Canadians (e.g., Bocking 1972; Holm 1988; Pentland and Hurley 2007), but this apprehension is never greater than when the water in question comes from the Great Lakes. Indeed, when the issue of exporting Great Lakes water has been raised, the Canadian public tends to consider it a largely *Canadian* issue. For example, in 1998, the government of Ontario issued a license to remove water by tanker from Lake Superior for export to an unspecified destination in Asia. The ensuing outcry was indicative of the importance of the Great Lakes – and of water in general – to Canadian identity. As Owen Saunders and Michael Wenig (2007) point out:

> Although the amount of water in question was not large, and although even a cursory analysis of the proposal would have made it plain that the costs involved were such as to make it virtually impossible for the scheme ever to go forward on a profitable basis, the very fact that such a license could be granted served to act as a touchstone for lingering concerns as to the security of Canada's water resources. (133)

Following publicity of this issue, Canada's federal government took action by attempting to negotiate a moratorium on water exports with each of the provinces and by amending federal legislation to preclude water exports from boundary waters. The government also acknowledged that this was more than just a Canadian issue: in 2000, it made a joint reference with its US counterpart to the International Joint Commission (IJC) to investigate and make recommendations on the broader question of water uses in the Great Lakes (International Joint Commission 2000). This incident reflects the degree of ambivalence that Canadians have with respect to the Great Lakes. On one hand, the lakes are thought of as a largely Canadian resource. On the other hand, Canadians know that management and protection of this resource requires a high degree of collaboration and cooperation with their neighbours to the south.

The Great Lakes in US History, Economy, and Culture

The citizens of the United States living within the Great Lakes watershed share their Canadian neighbours' sense of ownership over the lakes. Although the present era of Great Lakes may be characterized by cooperation fuelled by common attachment, the Great Lakes were perennially a theatre of conflict throughout the course of early US history. The common desire to share and protect this beautiful and irreplaceable natural landscape came about more recently, succeeding a past fraught with strife.

During the peace negotiations that brought the American Revolutionary War to a close, Benjamin Franklin suggested that Great Britain cede the entirety of the Great Lakes to the newly formed United States of America (Mansfield 1899, 124). When this proposal was rejected, the two sides eventually settled on a boundary line "through Lake Superior northward of the isles Royal and Philipeaux to the Long Lake" (Mansfield 1899, 124). The Great Lakes became the boundary between the new United States and British North America (Environmental Protection Agency 2011a).

The Great Lakes would again become central to conflict between the United States, Great Britain, and Canada during the War of 1812, in which one of the primary objectives of the US forces was expanding into and developing the area around the Great Lakes (Environmental Protection Agency 2011a). During the war, the lakes were the field of numerous battles and the site of a race for naval supremacy between the United States and Great Britain (Mansfield 1899, 154). When ambassadors from the United States and Great Britain met in Ghent in 1814, all of the diplomats had been instructed to negotiate a treaty that excluded the other side from the basin (Mansfield 1899, 181). However, the final treaty once again drew the boundary line through the midst of the Great Lakes (Mansfield 1899, 181).

Despite this bloody early history, the Great Lakes have inspired some of the United States' most enduring and significant political, cultural, and artistic works. The Northwest Ordinance of 1787, which numbers among the most important documents of the early United States (after the Declaration of Independence and the Constitution), set the standard for the admission of states into the union – a standard that was initially crafted for the Northwest Territories surrounding the Great Lakes (Duffey 1995, 929–30). The Northwest Ordinance prohibited slavery, provided for religious tolerance, and promoted education (Duffey 1995, 929–30).

The Great Lakes have also contributed to US culture as a source of inspiration to poets and other artists. During the region's early history, poets and explorers elegized the lakes' "sweet seas," "bold shores," and the "riches of the soil and the natural beauty of the country" (Sproule-Jones 2002, 21). The American poet Henry Wadsworth Longfellow immortalized the region, and Lake Superior in particular, in his *Song of Hiawatha*.

Some of the most ambitious engineering feats in the history of the United States have taken place in the Great Lakes Basin. The Mackinac Bridge, for example, is the third longest suspension bridge in the world. To construct the bridge (which opened in 1957), its engineers assembled the largest construction fleet that the world had ever seen (Mackinac Bridge Authority 2011). Between 1889 and 1910, by constructing a system of locks and canals, engineers reversed the flow of the Chicago River to end the dumping of untreated waste into Lake Michigan (Nilon 2004; Egan 2010). Despite unforeseen and significant environmental implications, the integration of Chicago's waterways to spur urban growth and trade on the Great Lakes was crucial to the historical development of this area as an industrial region (Cain 1998, 153).

In addition to being the site of numerous technological feats, the Great Lakes region fostered industry and the growth of urban centres within its watershed. Chicago grew because it sat astride the "gateway of commerce" (Cain 1998, 153). Detroit famously gave birth to the automotive industry when steamers carried steel and coal across the Great Lakes, and assembly lines churned out the automobiles that would so profoundly affect the US landscape and lifestyle across the continent (Henry Ford Museum 2003). Despite recent hardship in the Midwest, the Great Lakes continue to provide for the citizens of the Great Lakes states. A recent study by the Michigan Sea Grant at the University of Michigan found that 1.5 million jobs are directly connected to the Great Lakes, and $62 billion (USD) in wages are generated each year (Michigan Sea Grant 2011).

Consequently, US residents of Great Lakes states are fiercely protective of the water that shaped the entire history of their homeland, much like their Canadian neighbours to the north are protective of their history (Lasserre and Forest, in this volume). As described in detail in chapter 5, several proposals have been made to divert Great Lakes water to drier regions of North America. Americans and Canadians living within the Great Lakes watershed are strongly opposed to such proposals and have made protections against trans-basin diversions a priority in their cooperative water management policies.

The Great Lakes as a Site of International Cooperation

In his now classic article on conflict and cooperation along international waterways, Aaron Wolf affirmed that "shared interests along a waterway seem to consistently outweigh water's conflict-inducing characteristics" (Wolf 1998, 251).

The Great Lakes provide a good example of this rule, forming the site of a rich history of international cooperation over a shared waterway. Over the years, new and unforeseen developments in the political, technological, economic, and environmental sectors have presented themselves, requiring new forms of cooperation between the two countries to find resolution. We can trace this history of international cooperation as far back as the conclusion of the 1817 Rush-Bagot Agreement between Britain and the United States, which followed the hostilities of the War of 1812. This agreement drastically limited the naval force that each power could maintain on the lakes, effectively demilitarizing the space to allow for growth of international trade and facilitate commercial navigation and fishing.

A new suite of issues arose toward the end of the nineteenth century, including the question of financing capital improvements that would benefit both parties and concerns about the effects of diversions (such as that which took place in Chicago) on navigation, hydroelectric development, and water levels. Such issues gave rise to the creation of the International Waterways Commission in 1895 and were of central importance to the negotiation of the Boundary Waters Treaty of 1909 (BWT, or Treaty), which is discussed in detail below (Sproule-Jones 2002, 24). Around this time, technological developments, such as the invention of alternating current and improvements in the efficiency of turbines, allowed for the commercial exploitation of hydroelectricity, enabling the water of the Great Lakes, especially at Niagara Falls, to present itself as a new kind of resource. However, such developments also presented challenges that demanded international cooperation for resolution. For example, the prospect of both countries diverting water from the Niagara River to generate electricity produced an agreement by both parties "to limit the diversion of waters from the Niagara River so that the level of Lake Erie and the flow of the stream shall not be appreciably affected" (Boundary Waters Treaty, Article V). These provisions were later modified to accommodate changing circumstances. For instance, in 1941, allowance was made for increased hydropower generation on both sides of the border to help boost the war effort. A further joint revision in 1950 allowed for greater minimum flow over the falls, recognizing that protecting the "scenic beauty of this great heritage of the two countries" was the primary obligation of both countries (Wolf 1998, 260).

Opportunities for improved navigation of the Great Lakes have also provided an impulse for enhanced international cooperation. Proposals to enlarge and consolidate the separate canal systems in the Canadian and US sides of the border were discussed in the late nineteenth century. In 1921, the International Joint Commission recommended joint development of an enlarged navigation system on the upper St. Lawrence River. Finally, in the 1950s, parallel legislation passed by both federal governments allowed for construction of the St. Lawrence Seaway, a deep waterway that extends from the head of Lake Superior to the Atlantic Ocean (Easterbrook and Aitken 1956, 555).

Conservation and environmental concerns have also been a catalyst for international cooperation to protect the Great Lakes. Concerns about the depletion of commercial fish stocks helped give rise to the Inland Fisheries Agreement in 1908, through which Canada and the United States made an initial attempt to establish coordinated fisheries policies on the Great Lakes (Sproule-Jones 2002, 28). The invasion of the upper lakes by the parasitic sea lamprey in the mid-twentieth century, which resulted in decimation of commercial fish stocks, was the immediate reason for establishment of the binational Great Lakes Fisheries Commission in 1955. More broadly, this commission represents a joint effort to sustain commercial fisheries on the Great Lakes by helping inform coordinated regulations and fisheries policy (Sproule-Jones 2002, 28–9).

Water quality issues in the Great Lakes have prompted coordinated binational investigation and action to address emerging concerns over water pollution. In response to the overloading of the lower lakes, especially Lake Erie, by nutrients from urban wastewater and other sources in the 1960s, the first *Great Lakes Water Quality Agreement* was negotiated between the two countries in 1972. This agreement was highly successful in limiting inputs in the lakes of phosphorus – which had been identified as the most harmful nutrient – and was amended in 1978 to reflect emerging concerns about the effects in the basin of persistent toxic substances on ecosystem and human health. In 1987, Canada and the United States signed a protocol allowing for Remedial Action Plans to be put in place for 43 "Areas of Concern" around the Great Lakes that suffered particularly acute forms of pollution. The 1987 protocol also made provisions for "Lakewide Management Plans" to focus on critical pollutants and improve water quality in all five lakes.

Among the various water-related issues affecting the Great Lakes, the prospect of diverting water out of the basin has perhaps the greatest potential for dividing the interests of the two countries that share this resource. Here, Canada's longstanding concerns about exporting water come into play, since virtually all of the pressure – putative or real – to divert more water from the Great Lakes comes from south of the basin.[1] As Ralph Pentland and Adele Hurley (2007) have pointed out, "In the Great Lakes states ..., where population is more heavily concentrated than it is in Canada, inter-basin diversion possibilities are attracting the attention of several communities lying just outside or straddling the Great Lakes Basin divide" (166–7). Nevertheless, the threat of such diversions is felt by Americans living in the basin as well. Consequently, as with the other issues described in this section, the possibility of extra-basin diversions arguably has served to bring Canadians and Americans living in the basin closer together. The question is whether the combined interests of the basin will prevail over those outside the basin that might now, or in the future, covet water from the Great Lakes.

Along with threats to water quality, concerns about extra-basin diversions catalyzed the formation of Great Lakes United. This widely based environmental coalition is concerned with protecting the Great Lakes and St. Lawrence River ecosystem; it has offices on both sides of the border in Buffalo, Ottawa, and Montreal (Great Lakes United 2009). In 1985, the governors of the eight Great Lakes states and premiers of the two Canadian provinces signed the Great Lakes Charter, a nonbinding agreement that created a notice and consultation process for any significant new use, or increased diversion or consumption of Great Lakes water (Lasserre 2007, 156). Recognizing the need for a more effective mechanism, the *Great Lakes – St. Lawrence River Basin Sustainable Water Resources Agreement* (2005) was signed by the same parties. As described in greater detail below, its main provision bans exports and diversions of water out of the Great Lakes – St. Lawrence River Basin and prohibits new or increased withdrawals, unless strict criteria are met. Although still not legally binding, this agreement constitutes "a real political commitment to thwarting water export projects ... thus alleviating most concerns expressed by Canadians, specifically those living in Quebec and Ontario" (Lasserre 2007, 159).

The Current Situation – Policies and Governance

Managing Great Lakes water is necessarily an exercise in cooperation among multiple jurisdictions and levels of government. This transboundary challenge has produced a rich history of laws and policies that continues to develop today, and it demonstrates the cooperative culture in the Great Lakes region.

The Boundary Waters Treaty of 1909 and the International Joint Commission

As discussed in text boxes 1.1 and 1.2, the Boundary Waters Treaty of 1909 has foundational for the development of transboundary Canadian-US water management for more than a century. Since the region was relatively undeveloped until the late nineteenth century, there was little pressure on Great Lakes water resources and no need for international legal rules. By the turn of the century, both countries saw a need to avoid conflicts over use of the shared waters. The United States and Canada first established the International Waterways Commission in 1903 to address potentially conflicting rights in the countries' shared waterways (Woodward 1988, 326). The International Waterways Commission recommended that the two countries adopt legal principles of shared water use and form an international body to protect boundary waters. This recommendation led to the 1909 BWT, the first article of which provides for joint management and cooperation between the United States and Canada over boundary waters.

While navigation and access to boundary waters was the principle concern in 1909, the first draft included a provision forbidding water pollution with transboundary consequences, to be enforced by an international commission vested with "police powers" (Jordan 1971, 66–7). Thus, Article IV of the BWT provides: "... the waters herein defined as boundary waters and waters flowing across the boundary shall not be polluted on either side to the injury of health or property on the other." This provision establishes a clear standard regarding pollution of shared waters. Such pollution is just one form of transboundary water pollution, since contaminants often follow indirect paths of tributaries and different media, such as airborne pollution that is deposited into water bodies through precipitation. The underlying legal principle of Article IV, that one country's pollution should not harm another country, provides a foundation for Canadian-US international environmental law (Hall 2007).

The BWT also addresses the taking and diversion of boundary waters. Article III specifies that neither party may use or divert boundary waters "affecting the natural level or flow of boundary waters on the other side of the [border]" without the authority of the International Joint Commission (IJC, or the Commission). The IJC is a six-member investigative and adjudicative body, with the United States and Canada equally represented by political appointees. It is well respected in both countries and is often commended for its objectivity and leadership on environmental issues (Hall 2007, 706). This Commission's reports rely on the best available science and are free of nationalistic biases, making it an important source of information for the public and decision makers (Hall 2007, 707). Scores of issues have been referred to the IJC for nonbinding investigative reports and studies pursuant to Article IX. The Treaty only requires a reference from one of the countries to invoke this process, but as a matter of custom, this has always been done with the support of both countries (Hall 2007, 706–7). This bilateral approach has strengthened the credibility of the IJC's nonbinding reports and recommendations, and ensured sufficient funding for its efforts. These reports and their objective recommendations have enabled diplomatic resolution of numerous transboundary water disputes, as well as the crafting of new water-protection policies.

In recent decades, the Commission has played a critically important role in studying potential threats to the waters of the Great Lakes and informing both the public and decision makers in the United States and Canada. However, over the past several years, the role of the IJC in areas such as overseeing the *Great Lakes Water Quality Agreement* has been somewhat reduced (see below and Pentland, this volume). This Commission is also severely limited in its ultimate adjudicative power (Sadler 1986, 370–2). Together with the narrow scope of the Boundary Waters Treaty, this limitation has necessitated additional protections and joint-management programs for the shared water resources of the Great Lakes. We limit our discussion in the following sections to two of the most salient examples.

The *Great Lakes Water Quality Agreement*

In the 1960s, citizens and scientists became increasingly alarmed about water pollution in the Great Lakes. The United States and Canada therefore referred the pollution issue to the Commission in 1964. The IJC's 1970 report recommended new water-quality control programs and a new agreement on cooperation concerning pollution issues. In 1972, Prime Minister Pierre Trudeau and President Richard Nixon signed the *Great Lakes Water Quality Agreement* (GLWQA). This agreement recognized the grave deterioration of Great Lakes water quality, set forth general and specific water-quality objectives, provided for programs and other measures to help achieve these objectives, and redefined the powers, responsibilities, and functions of the IJC. Primary responsibility for implementation was left with the two federal governments, specifically the US Environmental Protection Agency and Environment Canada.

The 1972 agreement focused on phosphorous pollution. Sewage treatment was improved, and both countries adopted phosphate detergent bans. This success was tempered by new scientific discoveries and resulting public pressure to address persistent organic chemicals that "were already affecting the health of wildlife and could be a threat to human health" (Botts and Muldoon 2005, 27). The United States and Canada amended the GLWQA in 1978 (Article II) with a new purpose:

> [T]o restore and maintain the chemical, physical, and biological integrity of the waters of the Great Lakes Basin Ecosystem. In order to achieve this purpose ... it is the policy of the Parties that [t]he discharge of toxic substances in toxic amounts be prohibited and the discharge of any or all persistent toxic substances be virtually eliminated.

The parties signed another protocol in 1987 to add provisions for "Remedial Action Plans" for "Areas of Concern" and "Lakewide Management Plans" focusing on critical pollutants and drawing upon community involvement, focused on specific locations within the region. In 2006, the two countries and the IJC began conducting another comprehensive review of the GLWQA to address emerging threats to the health of the Great Lakes.

The GLWQA and its amendments have also given citizens an increased role in shaping pollution policy in the Great Lakes. Before the 1972 agreement, the Commission held public hearings on specific topics, but essentially conducted its business in private. Under increased citizen pressure about the environment, the GLWQA changed this custom and opened the Commission up to the public. The International Joint Commission (1998) affirmed its commitment to public participation in its *Ninth Biennial Report*: "The public's right and ability to participate in governmental processes and environmental decisions that affect it must be sustained and nurtured .... The Commission ... has come to expect, and to provide opportunities to be held publicly accountable for their work under the Agreement."

With increased public participation comes increased accountability for the two federal governments, and this informed and engaged citizenry has led to improved binational protection of the Great Lakes. An important element in public participation under this agreement is the Science Advisory Board, which is composed of scientists, citizens, and industry representatives. Originally called the Research Advisory Board, this body has a direct line of communication to advise the Commission. Despite its name, the Science Advisory Board has not limited itself to technical matters, and its work has led to many policy accomplishments (Botts and Muldoon 2005, 184–8).

The Great Lakes – St. Lawrence River Basin Sustainable Water Resources Agreement and Great Lakes – St. Lawrence River Basin Water Resource Compact

Shortened to "The Great Lakes Agreement" (2005) and "The Great Lakes Compact" (2008), these developments represent an advance in substantive legal rules for water use and cooperative management among the states and provinces sharing the Great Lakes Basin. This section focuses on the Great Lakes Compact as a new model for interstate water management and the Great Lakes Agreement as a new model for sub-treaty international cooperation. The foundation of these interstate and international management structures is the development of common standards for new or increased water withdrawals. These standards, which apply on both sides of the border, ensure the following: that water is used within the watershed, that individual and cumulative adverse environmental impacts are prevented, and that water use is reasonable and incorporates water-conservation measures.

The Great Lakes Compact establishes basin-wide common standards for water use and is recognized as an effective procedure for citizen participation on both sides of the border. Nevertheless, for constitutional and political reasons, this compact only includes the US states. Given the need for co-management of the Great Lakes as discussed above, state-provincial cooperation has been a regional goal for decades, but raises fundamental legal and political concerns. The Compact Clause of the US Constitution prohibits states from entering into a "treaty, alliance, or confederation" with a foreign government. In an attempt to meet the goal of state-provincial cooperation without running afoul of constitutional treaty limitations, the Great Lakes governors and premiers developed the Great Lakes Agreement (2005) as a nonbinding, good faith agreement that includes the provinces of Ontario and Quebec. This dual structure creates a legally and politically acceptable mechanism for cooperation with Canadian provinces.

The Great Lakes Compact incorporates the provinces through the Great Lakes Agreement's "Regional Body," which includes representatives from each state and province charged with conducting the "Regional Review" procedure. Although the Regional Body's authority is procedural rather than substantive, all parties consider it effective because it gives Canadian provinces and Canadian citizens a clear procedure for participating in major water-use decisions. The Regional Review process avoids infringing on federal treaty powers, but still gives the provinces an evaluative and procedural role that may prove useful for them. As noted above, despite the Great Lakes Agreement's nonbinding status, Canadian officials and environmental agencies are generally pleased with the Great Lakes Compact and the Great Lakes Agreement. Canadians are primarily concerned that the United States, with significant population growth in regions such as the south and southwest that are far from the Great Lakes Basin, will look to divert Great Lakes water to other parts of the country. Thus, Canadians welcome any legal limitations of Great Lakes diversions within the United States.

Looking Forward – Challenges and Opportunities

While the Great Lakes region has a relative abundance of water, as well as numerous policies and institutions for cooperative protection and management, some significant challenges loom on the horizon. As Pentland outlines in chapter 6, climate change, invasive species, energy development, and water pollution appear to be the most significant environmental challenges. Further, there is strong opposition to water commoditization and exports on both sides of the border. Given the region's strong history of cooperation, these challenges will likely be met with new agreements, policies, and institutions that build on the political, legal, and cultural foundation already established.

Climate Change

Climate change is expected to stress water resources and human communities globally, and the Great Lakes region won't be immune to these changes. While water levels in the Great Lakes have always fluctuated, the changes in levels have not been radical. Naturally fluctuating lake levels are critical for ecosystem function, but can be disruptive to human needs and economic development. Most climate models predict that water levels in the Great Lakes will drop during the next century, below historically fluctuating lows, by as much as 1.38 metres in Lake Michigan and Lake Huron due to changing precipitation, as well as increased air temperature and evapotranspiration (Hall 2010a, 249).

Scientists also expect air and water temperatures in the Great Lakes region to rise two to four degrees (C) by the end of the century. Lower lake levels and rising air and water temperatures will significantly impact fisheries, wildlife, wetlands, shoreline habitat, and water quality in the Great Lakes region. Economically, tourism and shipping are critically important to the region, and both industries are extremely vulnerable to the impacts that climate change is likely to have on water resources.

The increased variability in timing, intensity, and duration of precipitation will likely lead to increased frequency of droughts and floods in the Great Lakes region. The IJC estimates that stream runoff is expected to decrease, and baseflow (the contribution of groundwater to streamflow) could drop by nearly 20 per cent by 2030 (International Joint Commission 2003, 45).

Managing the Great Lakes under the conditions of hydrological uncertainty associated with climate change will pose a definite challenge for current agreements and processes of governance. However, the aforementioned Great Lakes Compact could be an ideal policy to help the region adapt to these conditions. While climate change will negatively impact the Great Lakes through deterioration of water quality, habitat, shorelines, and fisheries, as well as lower lake levels, the total available water supply will not be drastically reduced. The region's population and water usage are also not increasing significantly (in fact, in some locations and sectors they are actually decreasing), and freshwater is relatively abundant. The Great Lakes Compact is the most modern interstate water compact and was developed in recognition of the risks of climate change. It does not rely on fixed estimates of water supply to make definitive allocations; instead, it ensures sustainable water use by requiring states to comprehensively regulate water use to meet water conservation, ecosystem protection, and other standards.

Energy Development

Energy development in the Great Lakes also presents risks to freshwater. The Great Lakes have significant oil and gas resources that are economically and technologically accessible with modern drilling techniques. In a 2006 study, the United States Geological Survey (USGS) estimated that the US portion of the Great Lakes contains 312 million barrels of undiscovered, technically recoverable oil, as well as 5.2 trillion cubic feet of natural gas (Coleman 2006, 1–4). There are no comprehensive studies or estimates of oil and gas resources in this region under Canadian jurisdiction. The best information available is from the Ontario Ministry of Natural Resources, which estimates that the province's portion of the Great Lakes contains approximately 153 million barrels of recoverable oil and 1.5 trillion cubic feet of natural gas. Oil and gas production on the Canadian side of the Great Lakes dates back almost a century, with commercially produced natural gas taken from under the bed of Lake Erie as early as 1913.

Drilling for these oil and gas resources would create risks and potential impacts for the freshwater of the Great Lakes. In 2005, the US Army Corps of Engineers released a report to Congress titled, "Known and Potential Environmental Effects of Oil and Gas Drilling Activity in the Great Lakes." The report summarized that oil drilling and infrastructure would potentially "directly impact fish and wildlife habitats by clearing land areas or disturbing lake bottoms," and "the visual intrusion of oil and gas developments could reduce the desirability of these areas for tourism and other recreational uses" (Department of the Army – US Army Corps of Engineers Chicago District 2005, E-2).

The United States federal government, and most US states, have recently banned oil and gas drilling in the Great Lakes due to the environmental risks associated with these activities (Hall 2010b, 309–10). Canada, however, has not yet banned drilling in the Great Lakes. Ontario, the only province with significant Great Lakes jurisdiction, allows offshore gas wells and directional drilling of oil wells in this region (Hall 2010b, 310). Consistent with the cooperative and proactive nature of environmental policy-making in the Great Lakes, there is already an effort to enlist the IJC to study the issue and make policy recommendations. In the summer of 2010, more than 20 members of the US house of representatives from the Great Lakes states sent a letter to President Barack Obama, Canadian Prime Minister Stephen Harper, and the International Joint Commission. The letter urged the federal governments, in coordination with the IJC, to "undertake a review of oil and gas drilling by Canada in the Great Lakes, particularly in regard to safety, environmental impact and oil spill response plans" (Hall 2010b, 312–13).

Water Pollution/Quality

Thanks in large part to the *Great Lakes Water Quality Agreement* and other binational initiatives described above, the Great Lakes have exhibited significant improvements in water quality since the 1970s. Reductions in nutrient loadings (especially phosphorus) and in contamination by persistent, bioaccumulative toxic substances have been particularly noteworthy. General reductions in phosphorus loadings, for example, have prevented the recurrence of eutrophication in Lake Erie, which became a cause célèbre in the late 1960s. Reductions in persistent toxic substances have been accompanied by reductions in contaminant levels found in fish and wildlife in the basin. These substances no longer limit the reproduction of fish, birds, and mammals as they did in previous decades. Nevertheless, the Great Lakes continue to receive toxic contaminants from a wide variety of sources, including municipal and industrial wastewater, air pollution, contaminated sediments, runoff, and groundwater. Atmospheric deposition of toxic compounds is expected to continue well into the future and appears to be concentrated in urban areas around the lakes (Environment Canada and the United States Environmental Protection Agency 2009).

Despite the general reductions in nutrient loading after the 1972 *Great Lakes Water Quality Agreement*, phosphorus input to the lakes remains a concern, and in some locations, appears to be increasing because of agricultural and urban runoff, among other factors. An increasing proportion of this pollutant is found in a dissolved form that feeds near-shore algal blooms (Environment Canada and the United States Environmental Protection Agency 2009). In addition, invasive zebra and quagga mussels impact this issue by clarifying the water column, which allows deeper penetration of sunlight. When combined with the impact of increased sunlight, the continuing presence of phosphorus is considered a major contributor to algal blooms of nuisance proportions in nearshore waters, especially in Lakes Michigan, Erie, and Ontario (Environment Canada and the United States Environmental Protection Agency 2009, 11, 13).

Increasing concentrations of several "substances of emerging concern" in the lakes have also been observed in recent years. These substances include flame retardants, plasticizers, pharmaceuticals, and personal-care products that make their way into the lakes. They are of concern because they may pose a risk to the health of fish, wildlife, and humans. Flame retardants known as polybrominated diphenyl ethers (PBDEs) are found in many consumer products and have recently been added to the list of contaminants monitored in fish in Canada and the United States (Environment Canada and the United States Environmental Protection Agency 2009, 4). Citizens on both sides of the border have an interest in ensuring that the current review of the *Great Lakes Water Quality*

Agreement will adequately address these emerging threats and that the binational citizens organization Great Lakes United actively informs and engages citizens of both countries on this issue.

Biological Stresses: Invasive Species

The 1996 State of the Lakes Ecosystem Conference hosted by the US Environmental Agency and Environment Canada identified invasive species as among the greatest threats facing nearshore waters in the Great Lakes. At that time, 166 documented nonindigenous invasive species were inhabiting the waters of the lakes. Between 1996 and 2008, 19 additional invasions were reported (Environment Canada and the United States Environmental Protection Agency 2009, 12). A nonindigenous species is considered invasive when it is shown to negatively impact ecosystem health. The devastation wrought by species like the sea lamprey and the zebra mussel, as well as the threat posed by menaces such as the Asian carp, are obvious. However, all invasive species, including the smallest microflora and fauna, can affect ecosystem processes in unpredictable and harmful ways. Nonnative species have been linked to various problems in the Great Lakes, including increases in fish and waterfowl diseases, excessive algal growth, and the decline of important species at the bottom of the aquatic food chain (Environment Canada and the United States Environmental Protection Agency 2009, 2–3). Shipping, particularly the exchange of ballast waters, is considered the main vector for the introduction of invasive species in the lakes, but canals, online purchase of aquatic plants, and the aquarium and fish-bait industries are also sources of this problem. According to the 2009 State of the Lakes Report, "the Great Lakes ecosystem has been, and will continue to be, extremely vulnerable to introductions of new invasive species because the region is a significant receptor of global trade and travel. The vulnerability of the ecosystem to invasive species is elevated by factors such as climate change, development and previous introductions."

Conclusions

Environment Canada's web page on the Great Lakes makes the following, rather stark statement:

> The sustainability of the Great Lakes ecosystem is threatened. The ecosystem continues to experience ongoing biological, physical and chemical stresses, as well as new and emerging challenges like invasive alien species, new chemical contaminants and the impacts of climate change. (Environment Canada 2011)

At the same time, the agency recognizes that people living in the basin of the Great Lakes, as well as the governmental and nongovernmental organizations that represent them, are rising to the challenges presented by these and other threats:

> Many governments, organizations, groups and individuals are contributing to the restoration and protection of the Great Lakes. Work is being done at the local, regional, lakewide and basinwide scales, and all of these efforts help to restore and protect the Great Lakes. There are many success stories to be told but there is still work to be done. (Environment Canada 2011)

In this chapter, we have shown how the Great Lakes have presented a series of emerging challenges over the years, the resolutions of which have required a high degree of cooperation and collaboration among people on both sides of the international border. As previously suggested, there is every reason to expect that more complex challenges will continue to present themselves and that these issues will demand ever-higher degrees of cooperation within the region and between the two countries. While disagreements sometimes emerge, overall the citizens and political leaders in the Great Lakes continue to address freshwater issues with a culture of transboundary cooperation that helps create a model of Canada-US freshwater governance.

NOTE

1 For over a century, water has been removed from the Great Lakes Basin via the so-called Chicago Diversion, a canal built in 1900 to reverse the flow of the Chicago River and move water from Lake Michigan into the Mississippi River Basin. At 91 cubic metres per second, the Chicago Diversion is the largest extra-basin diversion from the Great Lakes; however, it is more than compensated for by a diversion of water from the Hudson Bay drainage basin into Lake Superior at Long Lac and Ogoki in Northern Ontario (Sproule-Jones 2002, 29–30).

REFERENCES

Bentley, D.M.R. n.d. "Thomas Cary's *Abram's Plains* (1789) and its Preface." http://www.uwo.ca/english/canadianpoetry/cpjrn/vol05/bentley.htm. Accessed 20 February 2011.

Bocking, Richard C. 1972. *Canada's Water: For Sale?* Toronto: James Lewis and Samuel.

Botts, L., and P. Muldoon. 2005. *Evolution of the Great Lakes Water Quality Agreement.* East Lansing, MI: Michigan State University Press.

Cain, Louis. 1998. "A Canal and Its City: A Selective Business History of Chicago." *DePaul Business Law Journal* 11: 125–83.

Coleman, J.L. 2006. "Undiscovered Oil and Gas Resources Underlying the U.S. Portions of the Great Lakes." *U.S. Geological Survey, Fact Sheet 2006–3049*. http://pubs. usgs.gov/fs/2006/3049/fs2006-3049_8.5x11.pdf. Accessed 16 August 2011.

Department of the Army – U.S. Army Corps of Engineers Chicago District. 2005. "Known and Potential Environmental Effects of Oil and Gas Drilling Activity in the Great Lakes." http://www.greatlakeslaw.org/blog/files/USACE_Great_Lakes_Oil_ Drilling_2005.pdf. Accessed 28 March 2013.

Duffey, Dennis P. 1995. "The Northwest Ordinance as a Constitutional Document." *Columbia Law Review* 95 (4): 929–68. http://dx.doi.org/10.2307/1123211.

Easterbrook, W.T., and Hugh G.J. Aitken. 1956. *Canadian Economic History*. Toronto: The Macmillan Company of Canada Ltd.

Egan, Dan. 2010. "Mayor Daley Floats Reversing Chicago River." *Journal Sentinel*, 11 September. http://www.jsonline.com/news/wisconsin/102676594.html. Accessed 28 February 2011.

Environmental Protection Agency. 2011a. *Great Lakes: Environmental Atlas and Resource Book*. http://epa.gov/greatlakes/atlas/index.html. Accessed 16 August 2011.

Environmental Protection Agency. 2011b. *Great Lakes: Basic Information*. http://www. epa.gov/glnpo/basicinfo.html. Accessed 16 August 2011.

Environment Canada. 2001. "Water, Art, and the Canadian Identity: At the Water's Edge." http://www.ec.gc.ca/eau-water/default.asp?lang=en&n= 62D086E9–1. Accessed 20 February 2011.

Environment Canada. 2011. "Great Lakes." http://www.ec.gc.ca/grandslacs-greatlakes/ default.asp?lang=En&n=70283230-1. Accessed 16 August 2011.

Environment Canada and the United States Environmental Protection Agency. 2009. *State of the Great Lakes 2009: Highlights*. http://binational.net/solec/sogl2009/ sogl_2009_h_en.pdf. Accessed 17 August 2011.

Grady, Wayne. 2007. *The Great Lakes: The Natural History of a Changing Region*. Vancouver: Greystone Books.

Great Lakes-St. Lawrence River Basin Sustainable Water Resources Agreement. 2005. http://www.cglg.org/projects/water/docs/12-13-05/Great_Lakes-St_Lawrence_River_ Basin_Sustainable_Water_Resources_Agreement.pdf. Accessed 17 August 2011.

Great Lakes United. 2009. "About Us." http://www.glu.org/en/about. Accessed 17 August 2011.

Hall, N.D. 2007. "Transboundary Pollution: Harmonizing International and Domestic Law." *University of Michigan Journal of Law Reform* 40:681–746.

Hall, N.D. 2010a. "Interstate Water Compacts and Climate Change Adaptation." *Environmental & Energy Law & Policy Journal* 5: 237–324.

Hall, N.D. 2010b. "Oil and Freshwater Don't Mix: Transnational Regulation of Drilling in the Great Lakes." *Boston College Environmental Affairs Law Review* 38: 303–14.

Heidenreich, Conrad E., and J.V. Wright. 1987. "Plate 18: Population and Subsistence." In *Historical Atlas of Canada, Volume 1: From the Beginning to 1800*, ed. R. C. Harris. Toronto: University of Toronto Press.

Henry Ford Museum. 2003. *The Life of Henry Ford*. http://www.hfmgv.org/exhibits/hf/. Accessed 28 February 2011.

Holling, Holling Clancy. 1941. *Paddle to the Sea*. Boston: Houghton Mifflin Company.

Holm, Wendy, ed. 1988. *Water and Free Trade: The Mulroney Government's Agenda for Canada's Most Precious Resource*. Toronto: James Lorimer.

International Joint Commission. 1998. "Ninth Biennial Report on Great Lakes Water Quality – Perspective and Orientation." http://www.ijc.org/php/publications/html/9br/covere.html.

International Joint Commission. 2000. "Protection of the Waters of the Great Lakes." Final Report to the Governments of Canada and the United States, 22 February.

International Joint Commission. 2003. "Climate Change and Water Quality in the Great Lakes Region." http://www.ijc.org/php/publications/html/climate/index.html

International Joint Commission. 2006. "A Guide to the Great Lakes Water Quality Agreement." http://www.ijc.org/en/activities/consultations/glwqa/guide2bw.pdf.

Jordan, F.J.E. 1971. "Great Lakes Pollution: A Framework for Action." *Ottawa Law Review* 5:65–83.

Lasserre, Frederic. 2007. "Drawers of Water: Water Diversions in Canada and Beyond." In *EauCanada: The Future of Canada's Water*, ed. K. Bakker, 143–62. Vancouver: UBC Press.

Linton, Jamie. 2010. *What is Water? The History of a Modern Abstraction*. Vancouver: UBC Press.

Mackinac Bridge Authority. 2011. "About the Bridge." http://www.mackinacbridge.org/about-the-bridge-8/. Accessed 28 February 2011.

Mansfield, J.B. 1899. *History of the Great Lakes*. Chicago: J.H. Beers & Co.

Marsh, James H., ed. 1985. *The Canadian Encyclopedia*. Edmonton: Hurtig Publishers.

McClelland, Ted. 2008. *The Third Coast: Sailors, Strippers, Fishermen, Folksingers, Long-Haired Ojibway Painters, and God-Save-the-Queen Monarchists of the Great Lakes*. Chicago: Chicago Review Press.

Michigan Sea Grant. 2011. "Study: More Than 1.5 Million Jobs, $62 Billion in Wages Directly Tied to Great Lakes." http://ns.umich.edu/htdocs/releases/ story. php?id=8280. Accessed 28 February 2011.

Morrison, Neil F. 1930. *A Commercial and Economic Geography*. Toronto: The Ryerson Press.

Nilon, Charles. 2004. "The Chicago River." *Encyclopedia of Chicago*. http://www.encyclopedia.chicagohistory.org/pages/263.html. Accessed 11 February 2011.

Pentland, Ralph, and Adéle Hurley. 2007. "Thirsty Neighbours: A Century of Canada-U.S. Transboundary Water Governance." In *Eau Canada: The Future of Canada's Water*, ed. K. Bakker, 163–82. Vancouver: UBC Press.

Reeves, Howard W. 2011. "Water Availability and Use Pilot: A Multiscale Assessment in the U.S. Great Lakes Basin, Complete Report." *U.S. Geological Survey,* Professional Paper 1778: 105.

Sadler, Barry. 1986. "The Management of Canada-U.S. Boundary Waters: Retrospect and Prospect." *Natural Resources Journal* 26:359–72.

Saunders, J. Owen, and Michael M. Wenig. 2007. "Whose Water? Canadian and the Challenges of Jurisdictional Fragmentation." In *Eau Canada: The Future of Canada's Water*, ed. K. Bakker, 119–41. Vancouver: UBC Press.

Sproule-Jones, Mark. 2002. *Restoration of the Great Lakes: Promises, Practices, Performances.* Vancouver: UBC Press.

Wolf, Aaron T. 1998. "Conflict and Cooperation Along International Waterways." *Water Policy* 1 (2): 251–65. http://dx.doi.org/10.1016/S1366-7017(98)00019-1.

Woodward, J. 1988. "International Pollution Control: The United States and Canada – The International Joint Commission." *New York Law School Journal of International and Comparative Law* 9: 325–44.

Looking Back, Looking Forward

12 Conclusion

ALICE COHEN, EMMA S. NORMAN, AND KAREN BAKKER

This volume has sought to introduce readers to some of the key trends and issues shaping water governance along the Canada-US border. The chapters answered the following three questions:

1. What are the most important trends in Canada-US transboundary water governance, focusing on the past century?
2. What are the key challenges and opportunities facing water managers, politicians, and water users?
3. How can Canada and the United States move forward with effective transboundary water governance?

This concluding chapter offers some final thoughts on each of these three questions in turn, below.

What Are the Key Trends in Canada-US Transboundary Water Governance?

Five trends are shaping (and reshaping) water governance along the Canada-US international boundary. The first trend is the changing role of federal governments with respect to transboundary waters: in general, a lower degree of federal government involvement is observed both in the United States and Canada. This trend is apparent in several examples: declining IJC references, the ongoing development of regional – as opposed to binational – transboundary initiatives, and greater engagement with local actors through, for example, the IJC's International Watersheds Initiative. Despite calls for greater federal involvement in a range of water management issues (e.g., Hill et al. 2008; Morris et al. 2007; Weibust 2009), this trend is likely to continue due to decreased funding for environmental initiatives and increased support for public and local participation in environmental decision-making. Indeed, the work presented in this volume suggests that federal governments tend to get involved once an issue is already "hot," rather than as a steady, ongoing presence. The exception to this pattern is, of course, the IJC, but the decline in references and reduced federal interest bring the relevance of this institution into question. These concurrent shifts will likely put significant pressures on local actors to fill the gap left by this new federal approach.

Related to this is a second trend: the increasing prominence of Aboriginal actors as sovereign. On both sides of the border, the increasingly important role of Aboriginal governments in formal decision-making processes is changing the landscape of Canada-US water governance. The Boundary Waters Treaty of 1909 (BWT, or the Treaty), which neglects to mention Aboriginal peoples, was not designed to incorporate greater recognition of Aboriginal self-government, treaty rights, and land claims. The question of how binational water governance arrangements will be shaped by this new dynamic is, as yet, unclear, but chapters presented here point to a number of examples – for example, Phare's chapter 2 highlights how First Nations are increasingly realigning water governance to reflect community priorities. This is also the case in the celebrated Yukon River Inter-tribal Watershed Council, which has won numerous awards for its ability to align governance practices with education, environmental protection, public health, and policy (see box 3.1). Furthermore, the St. Mary and Milk Rivers agreement highlighted in chapter 8 highlights efforts to optimize the existing governing arrangements to include avenues for greater tribal participation.

A third trend is the increasing importance of new management paradigms that incorporate ecosystem management and environmental objectives. Given increasing public concern over the types of environmental issues that Pentland raised in chapter 6 – most notably climate change and the emergence of new pollutants – it is not particularly surprising that an increasing number of North American jurisdictions are concerned with environmental protection. Yet this trend raises important questions about the relevance of the Treaty – which does not include provisions for many contemporary environmental issues – as well as the ability of binational frameworks to address the multiscalar nature of the origins and impacts of pollutants, as well as the effects of climate change.

The fourth trend relates to the evolving set of actors working with boundary and transboundary waters, conventionally the purview of national governments. An increasing number of nongovernmental actors are implicating themselves in the governance of waters located in, or flowing through, their areas of interest. In Chapter 11, Linton and Hall provided a number of examples of subnational cooperative organizations working across the Canada-US border in the Great Lakes – St. Lawrence region, and in chapter 2, Phare emphasizes the importance of Aboriginal actors in governing waters that cross international boundaries. Moreover, the development of regional identities that span national boundaries (such as the "Cascadia" or the "Salish Sea" in the coastal Pacific and the "Freshwater Nation" in the Great Lakes) shifts the focus from national identities to regional identities, at least discursively. Whether this shift translates to greater capacity for subnational actors to participate in transnational activities is yet to be determined. However, these regional identities can lead to cross-border influence – as in the Flathead case (chapter 10) – and cooperation – as in the Columbia case (chapter 11) – without necessarily involving federal government actors and binational institutions.

Finally, the chapters presented in this volume highlight the importance of considering the historical context in which specific transboundary arrangements were crafted, and, more broadly, the context in which the 1909 Treaty was written. At that time, environmentalism was largely focused on wilderness preservation and the creation – and contested commodification – of national parks (as symbolized by the work of John Muir, J. B. Harkin, and other early twentieth century environmentalists); in this context, water quantity took precedence over (arguably less well understood) water-quality issues. Moreover, the context in which the BWT was written led to the omission of explicit environmental protections in the treaty itself. This, in turn, reflects a broader challenge: we are addressing today's problems with a framework that is more than a century old. Indeed, the theme of Aboriginal sovereignty presents another case of a contemporary issue that is entirely absent from the dominant transboundary water governance framework.

These trends do not operate in isolation; they are interrelated and dynamic in nature and impact one another in various ways. For example, the growing importance of environmental issues raises two sets of questions with respect to Canada-US transboundary waters. First, how does this trend align with increasing Aboriginal self-government and treaty rights? We would argue that, taken together, these two trends present a powerful shift in transboundary relations: as the chapter on Devils Lake (chapter 9) demonstrates, environmental concerns are increasingly at the heart of transboundary water disputes. This is particularly true when environmental factors affect (or have the potential to affect) Aboriginal Treaty rights. Second, this prompts us to revisit the question of the relevance of the Treaty in the context of these two trends (which are not accounted for in its provisions). Indeed, considering the historical context can help illuminate what the Treaty does – and perhaps more importantly, does not – include. As discussed elsewhere in this volume, the BWT was drafted in a very different world, at a time when indigenous sovereignty and environmental protection were not on the radars of decision-makers. These issues are increasing in urgency and in public concern, and the

BWT will need to address these in some substantive way if it is to stay relevant in the coming decades.

What Are the Key Challenges and Opportunities?

The chapters in this book highlight a number of challenges and opportunities associated with Canada-US transboundary water governance.

One challenge relates to uncertainty, which we define as the unpredictability of factors impacting water governance. These uncertainties include – but are certainly not limited to – the emergence of new pollutants, the effects of climate change, uncertain economic fortunes and subsequent funding to binational organizations, the resolution of unresolved Aboriginal land and treaty rights, and the ever-changing Canada-US relationship more broadly. The upcoming renegotiation of the Columbia River Treaty adds a further layer of institutional uncertainty, which will be watched closely in the coming years as observers seek to understand how actors will deal with the thorny and the previously neglected questions of environmental issues and Aboriginal rights in the case of transboundary waters. The existence of uncertainty is, of course, not new, but the simultaneous changes presented by, on one hand, more flexibility, information sharing, and cross-border collaboration, and, on the other, the trend toward a "thickening" of the border, compound existing water governance challenges.

The juxtaposition of local innovations and rigid institutions is also reflected in a second challenge emerging from the chapters: that innovations at the border are framed by antiquated institutions – the century-old Boundary Waters Treaty, the impermeable and relatively unproblematized international border that privileges nation-states, and land-use priorities reflecting settlement patterns of a different era. As such, local-to-local groups are forced to work around, rather than through, these antiquated frameworks. This is not to say that no institutional flexibility exists (indeed, Clamen's discussion of the IJC in chapter 4 provides an example of this flexibility), but rather, that it is the exception rather than the rule.

Finally, and perhaps of greatest concern, is the challenge related to what we would argue is misplaced public focus. On the Canadian side of the border, we suggest that this misplaced focus takes the form of a deep rooted concern. Public concern with water exports is a highly emotive issue to Canadians. Whenever the spectre of water exports south of the border is raised, Canadians have responded with overwhelming negativity and concern over the export of the country's most precious resource. As explored by Lasserre (see chapter 5), small-scale water transfers are already underway. On the US side of the border, the public focus is often on water scarcity in the southern states and on transboundary relations between the United States and Mexico. This southern focus has perhaps come at a cost of greater public engagement with respect to the United States' neighbour to the north. We contend that this public concern is somewhat of a red herring, drawing much-needed attention away from the pressing issues identified above. Refocusing public attention (and the attendant political will it can generate) on questions of Aboriginal land and treaty rights, emerging new pollutants, and the effects of climate change – on both sides of the border – is both a challenge and an opportunity.

The chapters in this volume also explore a number of opportunities for improving Canada-US transboundary water governance. One important issue is the nature, scope, and legitimacy assigned to public participation – articulated with public attention to water issues and political will (as noted above). Participatory engagement ties in closely with many of the themes from this book, most notably the scaling "out" of decision-making processes from their state-centric predecessors to nonstate (and often local) actors. At the national scale, this trend is perhaps best exemplified by the IJC's International Watersheds Initiative, which seeks to engage the public in its programming. This move toward greater participation is distinct from the growing movement of Aboriginal self-government, which is grounded in law; Aboriginal groups are "not just another stakeholder" (NAFA 2000) but indeed are levels of government unto themselves. As several contributors have argued, these trends toward public participation present significant opportunity for moving ahead, because the unique expertise provided by nongovernmental experts and Aboriginal groups cannot help but enrich current understandings of transboundary waters.

In addition, another opportunity is reflected in the willingness for formalized governmental institutions to become more flexible in working toward effective transboundary water governance. The Commission's move to develop International Watershed Organizations is one example of this; the IJC responded to the international shift toward more holistic, watershed-scale planning and sought to fit it into its existing mandate. Moreover, as Norman and Bakker discuss in chapter 3, regional organizations such as the British Columbia – Washington Environmental Cooperation Council have established formal working roles, while endeavouring to address regional issues outside of the Ottawa-Washington framework. Outside of governmental institutions, the willingness of key actors to work both with and around centralized institutions has led to beneficial arrangements. In the Great Lakes, for example, a significant number of organizations are established as regional organizations working with, but distinct from, their federal government counterparts.

A third opportunity rests with federal governments in both countries, where, some have argued, there is significant room to expand the scope of activities while remaining well within the boundaries of the constitutional division of powers. In Canada, the federal government has been criticized for taking a "narrow view of its own powers" (Harrison 1996, 54) and "has often avoided exercising environmental authority it so clearly does possess" (Parson 2001, 6). This has resulted in a federal ceding of powers to the provinces in cases that are perhaps best addressed nationally. This has been the case with drinking water, for example, where some have argued that the significant diversity in drinking water standards across Canada's provinces is cause for concern (Hill et al. 2008).

Similarly, in the United States, there is significant room for greater federal action. The United States has long been lauded for its strong legislation such as the Clean Water Act (CWA), which eliminates toxic substance pollution into surface water and provides guidelines for water quality with the 1987 Water Quality Act amendments. However, concern has arisen in recent years that the integrity of the Clean Water Act is being compromised by lax enforcement and creative re-interpretations of water in the courtroom (Copeland 2010). For example, the CWA protects all waters with a "significant nexus" to "navigable waters." The interpretation of "waters" had over the years been interpreted as wetlands, intermittent steams, playa lakes, and sloughs. However the 2006 Supreme Court case *Rapanos v United States* held with a plurality,[1] viewing the term "waters in the United States" to include "only those relatively permanent, stranding or continually flowing bodies of water 'forming geographic features' that are descried in ordinary parlance as streams, ... oceans, rivers, [and] lakes." Under this ruling, the question of whether wetlands will be considered part of the CWA is open to interpretation. Even within this climate of legal uncertainty, however, significant opportunity exists for actors to work within existing frameworks to advocate for a stronger government action. This opportunity is particularly noteworthy in the case of local and nongovernmental actors who can work to apply pressure on governments to create meaningful public collaborations through a variety of "translocal" mechanisms (Conca 2008; Forest 2011; Karkkainen 2004). In parallel with the Canadian case, issues related to fragmentation are also cause for concern in the United States, where a number of agencies – including the Environmental Protection Agency and the US Army Corps of Engineers – have different mandates and goals (Conca 2008).

Although the inclusion of extra-governmental actors is important, and although Aboriginal, provincial, state, and regional organizations have a crucial role to play, there remains a critical role for the federal governments in overseeing transboundary water relations. At the very least, funding data collection and monitoring and enforcing existing agreements are the "low hanging fruit" of federal government involvement: activities that do not step on provincial or state toes but that would be beneficial to transboundary water governance. More controversially, federal governments could take on larger roles with respect to initiating reviews of existing governance arrangements and disputes (notably, although not solely, through references to the IJC, as Brandson and Hearne explored in chapter 9).

How Can Canada and the United States Move Forward with Effective Transboundary Water Governance?

We suggest that the answer to this question lies in the opportunities identified above: greater public participation, institutional flexibility, and federal involvement are all cornerstones of successful transboundary initiatives in the future. We are careful, however, to hedge our recommendations by outlining below some additional considerations.

The question of public participation is a knotty one, with scholars offering various competing perspectives. The dominant narrative in the policy literature is that the turn toward public participation in environmental decision-making is a positive development that can lead to better environmental outcomes, more locally appropriate policies, and greater public buy-in; moreover, public participation can legitimize processes that centralized governments conventionally have almost exclusively led in a top-down, command-and-control manner of decision-making (Beierle and Cayford 2002; Sabatier et al. 2005). We do not dispute the merit of these potential benefits, and we encourage Canada-US transboundary water institutions to incorporate public input into their decision-making processes. At the same time, however, it is important to be aware of the potential drawbacks of participatory processes. In addition to being more costly and time consuming (Nowlan and Bakker 2010), public participation holds two potential traps. The first is what Jesse Ribot calls the "charade" of participation, wherein power remains situated with government-led processes in provincial or national capitals, despite the decentralization and greater inclusiveness of decision-making processes. This phenomenon is reflected in chapter 3 of this volume, where authors Norman and Bakker outline the ongoing centralization of decision-making, despite the inclusion of an increasing number of extra-governmental actors at the border.

A second trap relates to the opposite problem: the withdrawal of government actors from decision-making processes, resulting in disorganization, disempowerment, and, perhaps most troublingly, the exacerbation of existing inequalities (Brown 2011; Marcus 2007; Reed 2007; Wilder and Lankao 2006). So where does this leave us in terms of our recommendation for increased public participation? We suggest that the uptake of participatory initiatives is a significant opportunity, when applied thoughtfully, resourced appropriately, and guided by clear goals and mandates, and when processes are inclusive, transparent, and include equitable representation from a wide variety of stakeholders and meaningful incorporation of public input into governmental decision-making processes. Moreover, we emphasize the importance of using local knowledge and expertise to supplement, rather than replace, more formal structures of transboundary decision-making since the existence of the border – which automatically draws in federal-level decision makers – is, for now, unavoidable.

We also suggest that existing institutions exercise flexibility and imagination in the application of their programs, including greater federal contributions to transboundary initiatives. We noted above that the IJC's International Watershed Initiative is one example of this of institutional flexibility, and there exist significant opportunities to expand this kind of thinking to other cases. Perhaps the most obvious of these opportunities is the greater inclusion of Aboriginal governments as "third sovereigns" in transboundary water initiatives, as well as a flexibility and receptiveness to the emergence of new pollutants and climate change impacts. In chapter 2, Phare highlights how the contributions of Indigenous communities in water governance processes are a critical step toward addressing social justice issues. Opening up space for engagement in a process is particularly important in a post-colonial context, such as North America.

The Boundary Waters Treaty was written in a specific socio-political climate; the question is whether it becomes obsolete or whether decision-makers (at all levels) can work with and around it to achieve better outcomes for people and the environment. We tentatively suggest that innovative solutions involve some combination of (1) stepping away from these dated treaties and provisions, and (2) that new transboundary agreements – such as the upcoming revision of the Columbia River Treaty – work to explicitly include the new considerations discussed throughout this volume in their provisions.

These discussions lead into a third point – one that we raised in the introductory chapter of this volume when we noted our support for the notion that lessons from the Canada-US example are transferable to other parts of the world. Although we believe this to be the case, we note that there is a subtle but important distinction between the transboundary *relationship* and the existing transboundary *institutions*. Several chapters in this volume have queried the relevance of the 100-year-old institutions, and we suggest that these merit careful consideration outside of the question of the relationship more generally. As chapters in this volume demonstrate, there are many examples of actors working across the border and developing innovative solutions and productive relationships outside of the formal institutional frameworks provided by the IJC and the BWT. The cases of the Great Lakes and the Columbia River show that – although certainly not flawless – the advancement of effective transboundary relations is not necessarily predicated on the existence of the formal institutions.

The Devils Lake (chapter 9) and Flathead (chapter 10) cases bring up a similar question through a different kind of example: what does it tell us that, in the most contentious examples, actors chose not to draw upon existing institutional frameworks? We thus pose a provocative question: is it the Canada-US relationship that is the world example, or the International Joint Commission and the Boundary Waters Treaty? Or is the former inextricably linked to the latter? And, given the number of recent examples circumventing the more formal and institutional channels of cooperation, might we look to other parts of the world for examples of effective transboundary water arrangements?

Indeed, some have suggested that the European Water Framework Directive (EUWFD) can be instructive for North America (and Canada, in particular) in exploring ways to put in place overarching coordinating mechanisms (Lagacé 2011). Moreover, through its use of river basis as the basic unit for measurement and management, the EUWFD provides an example of how water governance innovations can be built into the fabric of an agreement rather than layered on top. Furthermore, the Helsinki convention on Transboundary Rivers and Lakes (1992), developed under the auspices of the United Nations Economic Commission for Europe (UNECE) – has excellent practices with respect to monitoring, which could provide useful insights for North American transboundary waterways. Another example of governance innovation is the case of the Mara River Basin, which flows between Kenya and Tanzania and into Lake Victoria (the source of the Nile River). The basin has been highlighted internationally as an example of cooperation among local stakeholders, governmental officials, and external agencies (such as World Wildlife Fund), collaborating to address the challenges brought about by reduced flows, environmental degradation, and compromised water quality. The development of complementary national strategies in Kenya and Tanzania to support a transboundary water framework that includes participation in the decision-making process at multiple scales has been lauded as an important step in transboundary water governance in the global South (WWF 2012). The Mara River Basin may provide useful lessons on how to integrate participation at multiple scales into policy. Despite resource limitations, the governance structure provides meaningful avenues for community members to translate local knowledge into policy and adaptive management practices.

The chapters in this volume have shown a tremendous commitment to problem solving on the part of past and present federal, state, and provincial actors, as well as on the part of committed individuals and organizations working to find solutions to transboundary water challenges. Because of this commitment, the Canada-US transboundary water relationship has been held up as a preeminent example of how transboundary water governance ought to be done. Yet the volume has also pointed to a number of trends – emerging new pollutants, the effects of climate change, Aboriginal governments as a third sovereign, and declining references to our formal transboundary institutions – that are calling this preeminence into question. Will our existing institutions be able to cope with these dynamic changes, or will innovators need to seek new solutions that work around, rather than through, existing arrangements? We encourage readers to revisit the question posed by the title of this volume and ask whether – and under what conditions – the border is an obstacle to or facilitator of effective water governance. What governance strategies and institutional arrangements promote good water quality and thoughtful interventions, or (conversely) impede solutions and effective water management?

The Next 100 Years ...

Although in its most literal hydrologic sense, water is "borderless," the discussions in this book demonstrate the complexity of governing water in a world rife with political and cultural borders. Governing water with a recognition and appreciation of these multiple borders allows for a more nuanced approach to the increasingly complex and dynamic water issues facing the world today. For the foreseeable future, the governance of shared waters will likely require ongoing negotiations between these borders. However, we speculatively predict a paradigm shift that departs from the fragmented policies generated by geopolitical boundaries to strategic, integrated shared water governance models.

Our hope is that over the next 100 years of co-management and neighbourly relations, Canada and the United States will rise to the challenge of inclusiveness and equity; align policy more fully with science; integrate human health with ecosystem health concerns; and think beyond the bounded space of our own backyards, states, provinces, or countries. If we continue to work proactively, cooperatively, and in the spirit of water without borders, we have every confidence that these hopes can be realized.

NOTE

1 A plurality decision is different from a majority, as it is the greatest number of votes, but represents less than half of the votes.

REFERENCES

Beierle, T.C., and J. Cayford. 2002. *Democracy in Practice: Public Participation in Environmental Decisions*. RFF Press.

Brown, J. 2011. "Assuming Too Much? Participatory Water Resource Governance in South Africa." *Geographical Journal* 177 (2): 171–85. http://dx.doi.org/10.1111/j.1475-4959.2010.00378.x. Medline:21941692

Conca, K. 2008. "The United States and International Water Policy." *Journal of Environment & Development* 17 (3): 215–37. http://dx.doi.org/10.1177/1070496508319862.

Copeland, Claudia. 2010. "Clean Water Act: A Summary of the Law." Report No. RL30030. Congressional Research Service, Washington, D.C.

Forest, P. 2011. "Transferring Bulk Water between Canada and the United States: More Than a Century of Transboundary Inter-local Water Supplies." *Geoforum* 43 (1): 14–24.

Harrison, K. 1996. *Passing the Buck: Federalism and Canadian Environmental Policy*. Vancouver: UBC Press.

Hill, C., K. Furlong, K. Bakker, and A. Cohen. 2008. "Harmonization versus Subsidiarity in Water Governance: A Review of Water Governance and Legislation in the Canadian Provinces and Territories." *Canadian Water Resources Journal* 33 (4): 315–32. http://dx.doi.org/10.4296/cwrj3304315.

Karkkainen, B. 2004. "Post-sovereign Environmental Governance." *Global Environmental Politics* 4 (1): 72–96. http://dx.doi.org/10.1162/152638004773730220.

Lagacé, E. 2011. *Shared Water, One Framework: What Canada Can Learn from EU Water Governance*. Forum for Leadership on Water.

Marcus, R.R. 2007. "Where Community-based Water Resource Management Has Gone Too Far: Poverty and Disempowerment in Southern Madagascar." *Conservation & Society* 5 (2): 202–31.

Nowlan, L., and K. Bakker. 2010. *Practising Shared Water Governance in Canada: A Primer*. UBC Program on Water Governance. http://www.watergovernance.ca/wp-content/uploads/2010/08/PractisingSharedWaterGovernancePrimer_final1.pdf.

Parson, E. 2001. *Governing the Environment: Persistent Challenges, Uncertain Innovations*. University of Toronto Press.

Rapanos v. United States, 2006. Supreme Court of the United States 547 U.S. 715.

Reed, M.G. 2007. "Uneven Environmental Management: A Canadian Comparative Political Ecology." *Environment & Planning A* 39 (2): 320–38. http://dx.doi.org/10.1068/a38217.

Sabatier, P.A., W. Focht, M. Lubell, Z. Trachtenberg, A. Vedlitz, and M. Matlock. 2005. *Swimming Upstream: Collaborative Approaches to Watershed Management*. The MIT Press.

Wilder, M., and P.R. Lankao. 2006. "Paradoxes of Decentralization: Water Reform and Social Implications in Mexico." *World Development* 34 (11): 1977–95. http://dx.doi. org/10.1016/j.worlddev.2005.11.026.

WWF. 2012. Managing the Mara River in Kenya and Tanzania. Geneva: World Wildlife Fund Global. http://wwf.panda.org/who_we_are/wwf_offices/tanzania/index. cfm?uProjectID=9F0749.

List of Contributors

Karen Bakker is Professor, Canada Research Chair, and Founding Director of the Program on Water Governance at the University of British Columbia.

Nigel Bankes is Professor and Chair of Natural Resources in the Faculty of Law at the University of Calgary.

Elizabeth Bourget is an engineer at the Institute for Water Resources, US. Army Corps of Engineers.

Norman Brandson is Chair at Forum for Leadership on Water (FLOW) Canada.

Murray Clamen is Immediate past Secretary of the Canadian Section of the IJC, adjunct professor in the Department of Bioresource Engineering at McGill University, and a Member of FLOW.

Alice Cohen is an assistant professor in Environmental Science and Environmental and Sustainability Studies at Acadia University.

Noah Hall is associate professor of Law at Wayne State University.

Robert Hearne is an associate professor in the Department of Agribusiness and Applied Economics at North Dakota State University.

Frédéric Lasserre is Professor in the Department of Geography at the University of Laval and Project Director with ArcticNet.

Jamie Linton is Research Chair at University of Limoges on "capital environmental et gestion durable des cours d'eau".

Harvey Locke is a conservationist, writer, and photographer, and founder of the Yellowstone to Yukon Conservation Initiative.

Matthew McKinney is Director of the Center for Natural Resources and Environmental Policy at the University of Montana.

Emma S. Norman is an assistant professor of geography at Michigan Technological University. She holds a PhD from the University of British Columbia's Department of Geography and was a postdoctoral fellow with the Program on Water Governance.

Richard Paisley is Director of the Global Transboundary Water International Waters Research Initiative; Founding Director of the Dr. Andrew R. Thompson Program in Natural Resources Law and Policy; and an associated faculty member with the Fisheries Centre at the University of British Columbia.

Ralph Pentland is the President of Ralbet Enterprises Incorporated, a small-scale consulting firm; Acting Chairman of the Canadian Water Issues Council at the University of Toronto; and a Member of the Forum for Leadership on Water.

Merrell-Ann S. Phare is Executive Director and Legal Counsel to the Centre for Indigenous Environmental Resources

John Shurts is General Counsel of the Northwest Power and Conservational Council.

Index